旅居法國的亮亮老師親授

50道
法國阿嬤的
家常甜點

法國甜點
家中出爐

陳芋亮
Ines CHEN 編著

Pâtisserie
Chez Vous

橘子文化事業有限公司 出版

台灣人，法國魂

認識亮亮起於 13 年前的同事情誼，那時她是我們的造型師，以"龜毛"著稱，對於她的工作，她有她的專業跟堅持，以口才見長的我，常常被她的專業給說服。到了 2003 年的二月，她居然就這樣放下一切，在一句法文都不會說的情況下，隻身前往法國，本來以為她去遊學幾個月就會回來，沒想到她一待就是好幾年，等到再次與她相見時，她的身分已經從造型師，轉變為幸福美食的傳遞者。

還記得去年跟她合作"LIFE 樂生活"的第一集，我用大廚介紹她，她居然直接跟我說：「不要叫我大廚，我不是科班出身」，說實話，一般來賓是不會直接這樣糾正主持人的，但也因為這樣直來直往的個性，在一片電視料理節目的來賓中，她顯得特別鮮明與真實。

在節目上與書中的料理一樣，她示範的都不是我們所謂的「大菜」，必須穿西裝禮服拿刀叉才能享用的食物，每一道都是真正法國人家裡餐桌上會出現的。她用著法國人一貫的悠閒做著每一道料理，每一的步驟都像信手捻來般的不落窠臼，我都開玩笑的說她是「台灣人，法國魂」，老天讓她說著我們聽得懂的中文，傳達著法國人的生活態度。這本書，讓我這個料理白癡都有想買烤箱的衝動。一是有著豐富的圖片，將所有步驟圖像化，不用去猜想文字的意涵，二是畢竟很多食材都有地域性，法國隨手可得的材料，可能在亞洲見都沒見過，如果是買法國的翻譯食譜，這問題就會讓人很困擾，而亮亮在這部分也很用心，因地制宜，所以按圖操課，就算雖不中亦不遠矣。

希望這本書只是她的第一步，相信她從阿嬤身上學到的美食還有更多，在我把這本書的每一道料理都學完之前，她能把第二本寫好，我的小孩會很感謝她的。

禹安

Classic Desserts-Memories of Pastry Chef

Classic desserts are deeply rooted in our culture; they remind us of our childhood, of our after school snacks. The countless times we came back from school, hunting for that wooden spoon covered with melted chocolate from the mousse Maman just made; licking the spoons was like trophies! Then we would be looking much forward eating it after our meal, it was a daily ritual that probably etched our childhood memories forever.

Perhaps that, unconsciously, these sweets are the reason why I became pastry chef and baker in the first place; having a book that gather timeless classics from our childhood is a must have in anyone's library. What Ines Chen did with "*Pâtisserie Chez Vous*" is undoubtedly a must-have, sharing her great passion of timeless regional classics with everyone.

During my career, I have made very complex desserts, with textures, shapes, flavours, temperature and it has been wonderful. Yet, when I come home to my family, it always comes back to classics. I will mix a simple chocolate mousse for my son; bake a bunch of Madeleines for my wife and prepare other simple satisfying sweets to share with friends.

The importance of having recipes written and tested by someone who is not formally trained as a chef is a key point in connecting with the home audience, using gestures and words that people can identify with versus a too professional approach like it's often the case. It makes the recipes approachable and easy for everyone.

With Ines's book, anyone can bake these satisfying sweets in the comfort of their own home and why not, to create a tradition of their own with their family.

Gregoire Michaud

法式家常甜點：法式小確幸

在家製作美味的法式甜點，讓人人都能擁抱的一種簡單的幸福滋味，希望每位讀者都能夠享受這小小確定的幸福，感受法國阿嬤巧手所傳遞出的……

在法國生活就好比是鋼琴上不停跳躍的音符，是一個不停延續的樂章。開始接觸法國甜點跟家常料理，是在無心插柳的情況下；而做料理跟甜點，起初只是協助家裏的阿嬤[註一]準備家中家人每天享用的佳餚與餐後甜點，生活上，也跟所有的家庭主婦們一樣，每天忙碌着家裏大小事務。法國阿嬤巧手製作的甜點跟美味的家常佳餚，經常把家中的人餵養得非常飽足。法國鄉村家庭很重視全家人在餐桌上的用餐時光，餐桌上的時光伴隨着美味的食物以及讓人感到家庭幸福的甜點，餐間，家人肩並肩的聊天聲、酒杯碰撞聲、刀叉餐盤的摩擦聲、甚至是桌邊小朋友們的追逐遊戲聲，這每一個瞬間都源從法國美味的食物及甜點上桌開始，到結束用餐會不間斷地上演着，這就是法國鄉村家庭的小小幸福，餐後，大家期待的就是會讓人感到雀躍、開並且品嚐過程中嘴角不斷地上揚露出滿意且幸福的微笑，這就是我每天看似平凡卻又簡單的幸福。

最早從台灣來法國生活，單純地只為了這個美麗的語言"法文"而來到這個美食與甜點聞名世界的國家；幾年後，與現在的男友相識，進入這個略微龐大的家族，跟着法國家中的長輩在廚房裏學習法國家常料理，家常甜點，沒有任何專業烘焙料理背景的我，跟着阿嬤每天進出廚房，學習着沒有任何華麗裝飾、內在天然、外在樸實的每一道閃着幸福微光的好手藝。

幾年下來的法國家庭生活料理跟甜點學習後，我回到了我的家鄉——台灣，陸續開起了法式小餐館，提供着法國家庭的家常料理、家常甜點，精緻化的法國料理跟甜點充斥在台北的大街小巷，很少人知道什麼是真正的法國家常甜點跟真正的家常料理，我的咖啡館，基於我在法國生活的理念，提供着當天入庫的新鮮食材來製作料理，當天現做的法國家常甜點，愛嚐鮮的台灣人，慢慢地被我餐館裏實在味道以及沒有過多虛華包裝的外表所吸引，兩年後，淡出咖啡館經營，另外創立 Marmiton 廚房小學徒法式家常手工烘焙料理，在網路上銷售法國阿嬤真實、簡單味道的法國甜點，Marmiton 廚房小學徒有着我想要實踐法國家庭生活裏的自然、不造作、純粹、直接給予人幸福感的甜點。一開始接觸 Marmiton 廚房小學徒的朋友對我的甜點口感感到十分驚訝，原來法國的甜點可以品嚐到這麼豐富的味覺層次感。

註一：阿嬤是男友的媽媽，我跟隨小朋友稱呼她"阿嬤"，就如同阿嬤稱呼阿公作"爸爸"一樣。

Marmiton廚房小學徒提供訂購後才現做的甜點，傳遞給大家現做好可以現吃的概念。另外，也開課教授法國家常料理，家常甜點，目的在於讓大家知道法國家常料理真的一點也不難，有時候比中式料理還要簡單，除了吃的美味，極堅持法國鄉村生活方式，直接從自家菜園採蔬菜進廚房烹飪，自家種的蔬果最健康，除了有著為家人種植農作物時的那一份親力親為，每一口也都是凝聚家庭溫暖的力量，這樣的食材做起甜點及料理時特別地有感覺，也格外甜美。為了落實這樣的意念，在台灣，我親自拜訪有機小農，參觀果園，使用這些小農對自身栽種的農作物背後所付出的關心與細心照料的美味蔬果們，也希望能夠傳達給更多人如何料理，如何享用真正的食物味道。這樣的理念執行了快三年，Marmiton廚房小學徒的法國家常料理&甜點有著很好的成果，也因此讓我與許多有機小農們都成為不錯的好朋友。

今年，我回到法國，正式定居在法國，回到法國鄉村過著簡單的家庭庭園生活，在台灣沒有專業烘焙或是製作甜點背景的我，靠著阿嬤傳授的手藝，做出簡單的幸福甜點，我很幸運！我的生活裏有著法國阿嬤的陪伴，我將她的智慧與手藝跟我的實做經驗成甜點書，在這本甜點書裏，我以種類分類，並且以法國地區其代表性甜點的方式介紹法國甜點，然而，這些地區性甜點除代表著法國國內地區特色甜點之外，它們也已經深入在法國的每個家庭裏許久時間。在教授料理經驗及過程裏，知道許多朋友對看食譜有時會對書裏的步驟產生問號（畢竟沒有常做法式甜點的經驗，這是正常的現象）這本甜點書記載著許多我跟阿嬤學習時的小技巧與小祕訣，我盡量用圖佐文的方式讓製作過程更加清楚，大家練習時不會有模糊感，或許這本書會顛覆你以往對法式甜點的做法或是操作技巧，但別忘記了，這是法國家常甜點，不是專業烘焙食譜書，食譜裏很多都是法國阿嬤累積的經驗，道地家常味的甜點，又或是你擔心會做失敗？做失敗絕對不是你沒有天份，只是需要一點時間加以練習，讓你的手培養地更有手感，做起來的甜點成功率就會大大地增加，我不能說謊，說這書裏甜點都很容易上手，當然，書裏有許多步驟簡單的法式甜點，只需要4~5個步驟就可以完成的法式小甜點，同時也可以隨時製作好放在罐子裏慢慢地享用品嚐，有一部分的甜點需要花時間練習，並且按部就班的按照步驟來做，才能堆疊出成品的成功率；其實，我在學習甜點的路上也經常失敗（直到現在我嘗試做新甜點時，還是會算錯份量，做出來當然也就會失敗），但只要失敗，我便會馬上思考今天做的甜點那裏出問題？份量量錯？還是步驟不對？只要找出問題，下次再製作就會成功了！阿嬤也經常鼓勵我『失敗是走向成功的必經路』，每當我做失敗，她都會恭喜我耶，所以，不小心失敗的你，千萬別氣餒喔！

每個做甜點的人，必定難忘看到家人們或自己至親的友人品嚐自己親手做的甜點時，那種滿意、開心、幸福、說不出話的表情表現在臉上吧！看著我的甜點書，覺得外觀過於樸質嗎？打壞你對法國精緻甜點的美好印象嗎？真的，或許書裏這些甜點真的太過樸素，但是，相信我，現在翻起書選一道甜點，穿起圍裙，將烤箱加熱，帶著這本書一起進入廚房，製作一道甜點後，請你親愛的朋友或是家人們來品嚐，你將會在他們的臉部表情找到做下一道甜點的動力。

來吧！現在就拿著書，進入廚房，做一道讓家人們驚豔的法國家常小甜點吧！

目錄
Contents

part I

法國節慶糕點

聖誕節一過完，新年就不遠了。今年你可以改變與往常不同的方式來過新年，例如：跟法國人一樣做份國王餅，很有可能因此得到一年份的好運喔！

在法國每逢 1 月 6 日就是主顯節（L' Épiphanie），這一天不論是家庭聚會或好友相聚必定會吃國王餅，如果吃到國王餅裏面的陶瓷偶或是蠶豆的幸運兒要帶上皇冠接受大家的祝福，也代表你今年一整年會是非常好運一整年。

法國新年一月一整個月，大到賣場，小到超市，甜點店、麵包店開始銷售國王餅。因為國王餅在法國甜點上歷史相當悠久。許多的大型賣場裏銷售的國王餅大多都是以機器做大量製作。若想吃到好吃的國王餅一定要到 Boulanger 麵包店或是正統的 Pâtisserie 甜點店就可以買到純手工好吃的國王餅。

國王餅由兩張的千層派皮，裏面的餡料由杏仁粉、雞蛋、奶油、蘭姆酒混合而成的。進烤箱烤前的酥皮塗上蛋液後烤出來香脆的酥皮，在切的那一煞那間會聽到清脆的酥皮聲。如果沒有烤箱事先預熱是烤不出漂亮色澤的國王餅的。

溫溫熱熱的國王餅很好吃，冷着吃，味道可能就沒有那麼香，但是，能吃出香甜的甜杏仁餡的香氣與淡淡的蘭姆酒香。

關於國王餅（la galette des rois）有一個跟聖經有關的故事。根據聖經的記載，耶穌誕生後，三位有預言能力的聖賢，由遠方跟隨着天上最閃亮的一顆星星，即來到伯利恆找到藏在馬槽的瑪麗亞與聖嬰也就是耶穌。法國基督徒們深信這三位有預言能力的聖賢就是東方三國王。為了紀念東方三國王，法國在每年的 1 月 6 日開始準備甜的杏仁派，並在派裏放入一個小瓷偶（la fève），然後，由小朋友製作金色紙剪裁後的皇冠，接着搭配香檳或是氣泡酒一起享用。通常在週末的晚餐過後，由家中的長輩將甜杏仁派切成片，在蓋上一條好看的餐布或是餐巾紙，再將甜派轉一圈，由在場年紀最小的小朋友說個數字之後來分送甜杏仁派，並決定給誰開始先吃。誰吃到藏在甜派裏的小瓷偶，戴上皇冠就是這一天的國王，並接受大家的擁抱，親吻與真誠的祝福。

甚至法王路易十四的餐桌上也有國王餅。誰吃到蠶豆可以成為當天的國王，宮廷中的貴婦得到蠶豆當然就可以當一天的法國王后，並向真國王祈求恩寵，實現一個想要實現的願望。

國王餅
Galette des rois

食材 （約 6 人份）
千層派皮 500 克、杏仁粉 200 克、糖霜 125 克、室溫融化奶油 100 克、
蛋黃 2 顆、蘭姆酒 15 克、陶瓷偶 1 個

裝飾食材
蛋黃 1 顆、糖霜 25 克

Ingredients (Serves 6)

500 grams Puff Pastry

200 grams Almond Flour

125 grams Icing Sugar

100 grams Soften Butter

2 each Egg Yolks

5 grams Rum

1 each Porcelain Figurine

Ingredients for Decoration

1 each Egg Yolk

25 grams Icing Sugar

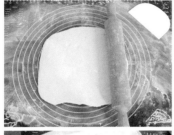

1

千層派皮分兩塊擀開約 5mm 厚度，兩張烘焙紙分別放在烤盤上下各放一塊擀開的派皮，放進冰箱冷藏，備用。

Divide pastry into two and roll out to 5 mm thickness. Use 2 pieces of parchment paper to separate the two pieces of pie dough on baking tray. Refrigerate for later use.

step 2

準備一個鋼鍋放進室溫融化奶油，用電動打蛋器將奶油打發。

Place butter into mixing bowl and whip with electric mixer until fluffy.

step 3

加入杏仁粉後，糖粉持續用打混合。

Add almond flour and icing sugar. Continue to mix until combined.

step 4

加入蛋＆蘭姆酒，將電動打蛋器增強，可以減少空氣進入餡料裏（意思也就是不要將空氣打進去），完全打混合後，使用矽膠刮棒將鍋邊的餡料全刮下攪拌混合。

Add in egg yolks and rum. Increase speed on electric mixer to reduce air entering into the fillings (meaning not to mix air into the filling). Once it is completely combined, use a silicone spatula to scrape all filling from sides of bowl to ensure even mixing.

step 5

烤箱以 200℃ 預熱 10 分鐘。從冰箱拿出派皮。擠上剛剛打好的杏仁餡，並將陶瓷放進去。

Preheat oven to 200℃ for 10 minutes. Remove pie dough from fridge. Pipe almond filling on top. Insert porcelain figurine into filling.

step 6

你可以將餡料裝入擠花嘴裏也可以直接將餡料塗在派皮上。派皮邊邊必須留約 2~3 公分的寬度。等一下做黏合作用。

You can either fill piping bag with almond filling and pipe it onto pie crust or spread it directly onto crust. Be sure to leave a 2-3 mm wide rim on the side of the crust to allow binding of the bottom and top crusts .

step 7

派皮預留的邊邊一圈塗上水，接著放上另一張派皮用手指將派皮壓黏合在一起。

Brush bottom crust rim with water. Lay second piece of pie crust on top. Use your fingers to carefully secure the two pieces of dough together.

step 8

塗上蛋黃水，放冰箱冰 15 分鐘。

Brush top of pie crust with beaten egg yolk. Refrigerate for 15 minutes.

step 9

拿出冰箱的派，使用刀背在派的外緣部刻畫上線條。另外用牙籤在派的頂端中間跟四邊戳洞，國王餅烤的時候會膨脹，戳洞為了方便餅裏的空氣出來，不會爆開。

Remove pie from fridge. Using the back of a knife, make slits on outer side of the pie. Using a toothpick, prick holes on the top center and 4 sides of the pie. The pie will expand while it is baking. Pricking holes will release the air inside the pie and prevent the crust from cracking.

step 10

用刀尖在國王餅面部劃上線條，或是皇冠上面的幾何圖形或是葉子的圖樣都是國王餅象徵的圖案。

Using the tip of a knife, draw lines, or shapes on the crown, or leaves. These are all symbols of the Galette des Rois.

step 11

進烤箱前撒上糖霜，200℃ 烤 10 分鐘，180℃ 烤 20 分鐘。

Sprinkle pie with icing sugar before baking it at 200℃ for 10 minutes. Then reduce temperature to 180℃ and bake for another 20 minutes.

step 12

出烤爐後，塗上糖水（不是必須的）。等餅冷了就可以跟全家來玩尋寶吃餅的遊戲了！

Once it is done, brush it with syrup (optional). Let it cool and then have the whole family enjoy this treasure hunt and pie eating game!

Helpful Tips
高亮的小建議

1. 兩張派皮在黏合時候，如果不一樣大小，可以用刀子修整出一樣的大小形狀。

2. 使用刀子畫圖案時，通常一開始會抓不準的手力道，可能會一下子下太大力將派皮割破，所以，每一刀都要輕輕割的下喔！

3. 烤前撒糖粉，烤後就不用在淋糖水了！撒糖粉為可以烤出漂亮且均勻色澤的派皮表面。若是烤完仍想淋糖水，可使用 25 克糖 50 克的熱水，將糖融化要放冷卻。之後，國王餅烤出來趁熱馬上淋上，餅的表面顏色會變漂亮焦糖色。

1. *If the two pastry dough are not even when binding together, use a small knife to trim them until they fit perfectly.*

2. *Be gentle when you are making designs with the small knife. If you are not careful, you may break the pie crust.*

3. *If you sprinkle crust with icing sugar before baking, it is not necessary to pour syrup on top after it is done! Icing sugar ensures the top to be baked nicely and evenly. If you still want to pour syrup on after it is done, dissolve 25 grams of sugar in 50 grams of hot water. Let syrup cool down before pouring onto pie. Otherwise the top of the pie will turn into a caramelized color.*

法國南邊在主顯節一月六日（L'Epiphanie）這期間才會出現的傳統蛋糕「奶油國王蛋糕」。奶油國王蛋糕是為紀念耶穌誕生（誕生於聖誕節）東方三國王前來探望小耶穌，贈送給耶穌的糕點。這是在南法 Languedoc 享譽盛名的一款蛋糕。蛋糕裏有着淡淡橙花水的香味，裝飾在蛋糕上面的糖晶跟糖漬水果代表着國王皇冠上面那些富麗堂皇、五顏六色的寶石。因此另有國王蛋糕（gâteau des rois）的語意。蛋糕裏面藏着一個陶瓷偶，按照習俗誰擁有這個隱藏在蛋糕裏的陶瓷誰就是當天的國王，他可以戴上紙做的金色皇冠或是自己親手做的皇冠，得到陶瓷偶與戴上皇冠的國王就要負責購買或是製作明年的奶油國王蛋糕。

16

奶油 國王蛋糕
Brioche de rois

Ingredients (Serves 6)

300 grams All Purpose Flour

25 grams Granulated Sugar

4 each Eggs

150 grams Soften Butter

5 grams Fresh yeast

100 ml Water

1/2 tsp Sea Salt

2 tbsp Rum

1 each Sweet Orange

120 grams Candied Fruits
(orange or lemon)

1 each Ceramic Figurine

Ingredients for Decoration

100 grams Candied Fruits

40 grams Pureed Sweet Orange or Marmalade

15 grams Crystal Sugar

1 each Egg Yolk

Gabulle in Wonderland

王冠紙樣可在此下載

食材（6 人份）

麵粉 300 克、細砂糖 25 克、蛋 4 顆、室溫融化奶油 150 克、
新鮮酵母菌 5 克、水 100 毫升、海鹽 1/2 茶匙、蘭姆酒 2 湯匙、
甜橙 1 顆、糖漬水果（甜橙片或是檸檬片）120 克、陶瓷 1 個

裝飾食材

糖漬水果 100 克、甜橙泥或是果醬 40 克、糖晶 15 克、蛋黃 1 顆

製作

 step.1

製作麵糊：先將甜橙洗乾淨，擦乾後將甜橙皮磨下來，備用。糖漬檸檬片切丁，備用。

Dough: Wash sweet orange and wipe dry. Scrape orange rind. Dice candied fruit. Set both aside.

 step.2

將新鮮酵母菌放進水裏，融化攪拌。

Add fresh yeast into water. Stir to dissolve.

step 3

在一個鋼鍋裏加入麵粉。麵粉挖一個洞加入細砂糖、海鹽、新鮮酵母菌,用手將所有材料完全揉合一起。

Add flour to stainless steel mixing bowl. Make a hole in the center of the flour and add sugar, sea salt and yeast. Mix lightly with hand until combined.

step 4

接着加入雞蛋、蘭姆酒、甜橙皮。再將麵團揉混約 5 分鐘,麵糊很黏稠,可以使用電動打蛋器慢慢打混合。

Add egg, rum, and orange rind to the flour mixture. Continue to mix dough for about 5 minutes. Dough will be very sticky. You can use an electric mixer to slowly combine them together.

step 5

再加入奶油,慢慢的打混麵糊。直到麵糊有點彈性即可。因為麵糊很黏稠,打到有彈性的時候用攪拌棒挖起時可以看到麵團有點彈性的彈回去。

Add in the butter and continue to mix the batter slowly until it is slightly elastic. Because the dough is very sticky, once it is elastic, you will notice that it will bounce back when stirred with spatula.

step 6

放入切丁檸檬丁跟陶匙攪拌後,蓋上一塊乾淨的布放進冰箱一晚。

Add in candied fruit and ceramic figurine. Cover with a piece of clean cloth and leave it in the refrigerator overnight.

 隔天 *The Next Day*

step 1

倒出麵團後,用手在揉約 2~3 分鐘。

Pour dough out and knead for about 2 to 3 minutes by hand.

step 2

使用擀麵棍將麵團擀開在摺疊 4 次。再擀開再摺疊 4 次。

Use rolling pin to roll dough out and fold it 4 times. Roll out again and fold again 4 times.

step 5

放置一個小時。

Let dough sit for an hour.

step 7

蛋黃加少量水混合後，塗在麵團上。

Add a little water to the egg yolk. Brush onto dough.

step 3

將麵團揉成圓形麵團，用手指戳進麵團中間，將洞挖大一點。

Shape dough into round ball. Stick finger into the middle of the dough and make a hole.

step 6

烤箱以 170℃ 預熱 10 分鐘。等到中間洞的麵團變硬後，再稍微移動一下麵團，讓麵團可以再度發酵。

Preheat oven at 170℃ for about 10 minutes. When the dough around the center hole becomes hard, move dough slightly and allow dough to rise again.

step 8

放進烤箱烤 45 分鐘。

Bake for 45 minutes.

step 4

中間挖洞後，麵團如果會縮會去，中間的洞找個圓形烤模或是烘焙紙折成圓圈後放在中間，避免中間的洞彈回變小。

If dough bounces back after hole was made, put a round cookie cutter or paper cylinder to retain the hole.

step 9

直到奶油蛋糕烤熟保持再微溫的狀態塗上甜橙泥或是果醬，在撒上糖晶跟自己喜歡的糖漬水果做裝飾。

After butter cake is done and still warm, spread pureed sweet orange or marmalade on top. Sprinkle with crystal sugar and your favorite candied fruits.

Helpful Tips
亮亮的小建議

1. 麵糊在做的時候，非常得黏稠，如果沒有專門攪麵團的機器，可以跟我一樣，用電動打蛋器慢慢打也可以打得很均勻，讓麵團有彈性。

2. 這個蛋糕較花時間製作，但是完成後絕對會非常有成就感喔！

3. 糖漬水果的部份沒有規定一定要用哪一種糖漬水果，只要家人喜歡的水果口味即可。

4. 若不喜歡口感太甜的朋友，在擺糖漬水果裝飾時，就不用在塗上糖水了。若是喜歡糖漬水果有點光澤建議用 100 克糖跟 50 克水煮糖水放涼後塗在糖漬水果上。這樣優點是糖漬水果不易掉落，也會增加賣相。

5. 如果可以的話可以加幾滴橙花水（Fleur d' orange），因為這個蛋糕的特色就是要有橙花香味。

6. 找不到蛋糕裏的陶瓷的話，可以用杏仁或是蠶豆來替代。

1. *Dough is very sticky. If you do not have counter top mixer, you can use a handheld mixer like I do and slowly mix the dough. Dough will still be elastic.*

2. *This is a time consuming cake. However, you will have a very satisfying sense of achievement once it is done.*

3. *There is no specific requirements as to what kind of candied fruits should be used. Any kind that you favour will do.*

4. *If you want the cake to be less sweet, you can skip the step of brushing sugar water on top of the candied fruits. If you want the candied fruits to have a glazed look, you can brush it with cooled 100 grams of sugar dissolved in 50 grams of water. This will prevent the candied fruits from falling off the cake and also give it a more presentable look.*

5. *You can also add a little bit of Fleur d'orange, as this will bring out the orange flavour of the cake.*

6. *If you cannot find the tiny ceramic figurine, you can replace it with almond or fava bean.*

蛋糕麵包
Pain-gâteau

蛋糕麵包又另一個名字 "Koekbrood"。

在北法，尤其靠近加萊海峽省（Pas-de-Calais），不論任何一種節慶日都都不會錯過品嚐這個糕點的好機會。例如：農作物收穫期，啤酒花採收期之後的第一個星期天的早上，在餐桌上會出現奶油、火腿跟冷湯，跟好大一塊的蛋糕麵包，讓大家大快朵頤的吃。或是聖誕節的早晨，幾片的蛋糕麵包，塗上奶油，在一杯加熱融化巧克力加入牛奶撒上大量的巧克力碎片一塊享用。

> 沒有任何一個節慶，這個鬆軟散發著有淡淡橙色花香味，像蛋糕又有麵包香氣的"蛋糕麵包"會在北法任何一個家庭的餐桌上缺席。
> 任何的節慶機會怎麼能夠錯過品嚐鬆軟的蛋糕麵包跟一杯濃郁現煮的熱巧克力呢？

Ingredients (Serves 4 to 6)

250 grams All Purpose Flour

2 each Eggs

100 grams Melted Butter

1 handful Raisins

1/2 large bowl Hot Tilia Blossom Tea

1 tbsp Granulated Sugar

41 grams Fresh Yeast

1/2 cup Milk

食材（約 4~6 人份）

麵粉 250 克、蛋 2 顆、融化奶油 100 克、
葡萄乾一小把、熱的菩提花茶半大碗、
細砂糖 1 湯匙、新鮮酵母菌 41 克、
牛奶半杯

製作

step 1

將葡萄乾放進菩提花茶裏，泡約
10 分鐘。

*Add raisins into hot tea . Let it steep for
10 minutes.*

step 3

蛋白用電動打蛋器打發，備用。

*Beat egg white with electric beater until
soft peaks formed.*

step 5

麵粉放進鍋裏中間挖出一個洞，
放進蛋黃；攪拌混合。

*Add flour to mixing bowl and make a hole
in the center. Add egg yolk and mix well.*

step 2

分開蛋黃跟蛋白。

Separate egg white and egg yolk.

step 4

糖與新鮮酵母菌放進牛奶裏攪拌
混合均勻。

*Add sugar and yeast into milk and mix
well.*

step 10

烤箱 180℃ 預熱 10 分鐘，烤 30~40 分鐘。

Preheat oven to 180℃ for 10 minutes. Bake dough for 30 to 40 minutes.

step 6

接着加入融化奶油跟酵母菌牛奶糖液，攪拌混合均勻。

Add melted butter into yeast and milk mixture. Stir well.

step 8

再加入打發蛋白慢慢攪拌跟麵團融合。

Gently fold beaten egg white into batter.

step 11

出爐後，冷了切片享用。

Cool, slice and enjoy!

step 7

葡萄乾從菩提花茶裏撈出瀝乾水份後加入，將麵糊跟葡萄乾攪拌混合。

Remove raisins from tea and drain. Add raisins into batter and mix well.

step 9

放進烤模裏，放在較溫暖的地方放置 2 小時。

Put batter into baking pan. Keep in warm place for 2 hours.

Helpful Tips
亮亮的小建議

菩提樹花也可以使用其他的花茶用熱水泡開後使用。

If you do not have Tilia Blossom Flower tea, you can replace it with other flower tea.

法式薄餅

crêpe 薄餅的來由歷史

薄餅不是在近幾年出現在法國的新創意甜點，它的存在經過了法國各方查證，尋找關於薄餅的來由，歷史與記錄。最後的記錄，薄餅最早出現在 7000 年前比耶穌還要在更早存在的一種麵食食物。在這段早期期間，法國尚未稱它為 "crêpe"。那時候薄餅的外觀比較像大而薄的烤餅。這種烤餅曾經是用一種麵糊混合物來調配而成。之後，變成人們將種植的穀物壓碎成粉末後注入水再調配成麵糊，這段時期是以熱平石板來替代平底鍋煎熟烤餅。不論是 crêpe 薄餅或是 galette 烤餅（俗稱 "鹹薄餅"）都是在 13 世紀時出現在法國的布列塔尼地區，由自亞洲傳入的蕎麥麵粉成為鹹薄餅主要的麵粉，更是這兒居民們日常生活飲食用麵食之一，白麵粉則是用來做一般雞蛋跟牛奶調配後而成的甜薄餅。

現今，薄餅在法國到處可見，像是大賣場，小流動餐車還是你想來個布列塔尼式的風格吃法，那就找一家當地人會去的薄餅店，點一盤薄餅和一杯當地特色的蘋果氣泡酒（Cidre）。法國家家戶戶的媽媽、奶奶們會製作這種十分容易準備，好製作法國家常甜點，薄餅可是所有法國人孩童時期快樂回憶的其中一種甜點。

聖蠟節 La Chandeleur

法國的聖蠟節 La Chandeleur 聖誕節過後的 40 天，確切的日期在 2 月 2 日這天。耶穌的信徒們在這大都會到自己當地的耶穌教堂做禱告跟祈福，教堂並會佈置聖母瑪麗亞生下耶穌的相關文獻，畫像供大家瞻仰。為什麼會是在 2 月 2 日呢？因為，這天正是聖母帶着出生後 40 天的小耶穌到主堂瞻禮的節日，因此，又稱做瞻禮節或是洗身禮。

在古老的法國時期，法文的二月寫法為 Février 即有洗淨、滌除的意思，剛好這個節日與即將離去的冬日淨化階段有些關係性，更隨着大地漸漸的白天在復甦，日光時間漸漸拉長。意味着冬天離去，迎接風光明媚的春、夏天。這天教堂裏都會有遊行隊伍每人手拿着一根蠟燭，另外教堂各角落將點上蠟燭，信眾們也會點上蠟燭並向耶穌祈福，同我們到廟宇裏點香供佛並向神明祈福求平安的意思。

二月正是很多鄉鎮地方開始進行各種歡樂節慶，吃吃喝喝狂歡的時候，因為跟着洗身禮到來與冬天離去，接着歡度迎接春，夏天來臨。

2 月 2 日這天，不論家中還是法國許多地方舉辦活動都要以 "crêpe 薄餅" 做活動主題，總之，無論如何這天就是吃薄餅就對啦！但，又為什麼要吃薄餅呢？阿嬤説，耶穌跟太陽一樣是帶給大家光明，希望。薄餅大又圓，煎成金黃色像是太陽。真好符合送走陰冷的冬天後，薄餅像是我們迎接暖暖的陽光般的到來。這樣説來有點頗像是我們中秋節賞月吃月餅的道理。

> 啤酒風味的薄餅是法國北邊與比利時一帶非常俱有特色的地區性法式薄餅，蛋奶口感的薄餅皮夾有着啤酒香。薄餅要好吃在煎的時候餅皮一定要沒有氣孔，手感軟嫩。法式薄餅吃法相當多元化，就如同吃鬆餅一樣，喜歡什麼醬料或是水果都可以添加。法國最簡單基本的甜薄餅吃法是薄餅撒上些許白糖，或是塗上果醬，放上各種莓果，新鮮水果，它可以是甜點，也可以是早餐或是晚餐。
>
> 我們一般吃薄餅習慣喝上一杯氣泡蘋果酒（Cidre）不論是甜味的蘋果酒（Doux）或是帶澀味的蘋果酒（Brut），吃薄餅搭配氣泡蘋果酒好有風味喔！如果是自家釀造的當然是再好不過了。

{ Ingredients for Beer Batter Crêpes

180 grams All Purpose Flour

55 grams Melted Butter

500 grams Milk

3 each Eggs

3 grams Sea salt

20 grams Granulated Sugar

1/2 can Golden Color Beer

啤酒風味薄餅皮材料

麵粉 180 克、融化奶油 55 克、
牛奶 500 克、蛋 3 顆、海鹽 3 克、
細砂糖 20 克、黃金色啤酒 1/2 罐

製作

step 1

麵粉，糖跟海鹽放進一個鋼鍋
裏，用手和勻。

*Add flour, sugar and sea salt in stainless
steel bowl. Mix well with hand.*

step 2

蛋事先打散，在麵粉中間挖出一
個洞，倒入蛋液，攪拌均勻。

*Beat egg. Make a hole in flour mixture
and add egg . Mix until combined.*

step 3

倒入一半牛奶慢慢攪拌與剛剛和
勻的蛋液麵團一起攪拌，直到牛
奶與麵團充分融合倒入剩下一半
牛奶，在倒入融化奶油跟啤酒。
攪拌後放置 30 分鐘或是 1 小時。

*Pour half of the milk into the batter .
After the milk is completely absorbed
by the batter, add the remaining milk,
melted butter and beer. Mix well and let
rest for 30 minutes to an hour.*

step 4

開中小火，平底鍋上抹上些許
奶油，待鍋子熱後，鍋子離火
加入 1 湯瓢的麵糊液。倒入麵
糊液時請從鍋邊倒入在轉動鍋
子讓麵糊液可以充分流動平均
在平底鍋裏。

*Over medium heat, add some butter to
frying pan. After pan is heated, take it
away from flame and add 1 soup ladle
of batter. Make sure you pour in batter
from side of the pan. Move pan slightly
so batter can be evenly distributed all
over the pan.*

step 5

煎到鍋邊的麵糊液熟成翹起來就
可以翻面，在煎個 1 分鐘就可以
挪到盤上了。一直從複剛剛的步
驟直到將麵糊都煎完。

*Once the batter on the side starts to
brown, flip crepe over and cook for
another 1 minute. Continue until all
batter is done.*

附上幾道簡單的薄餅甜點作品

傳統白糖薄餅

焦糖薄餅

part 2

法國地方性家常
特色塔派

大黃根
蛋白霜塔
La tarte meringueé à la rhubarbe

北部 加萊海峽區 Pas-de-Calais

大黃根是種很特殊並別具風味的根莖類蔬菜，在亞洲地區鮮少人使用或是知道如何使用它。大黃根除了做主菜旁的美味配菜，也經常性地的被使用在甜點上。大黃根酸酸的口感，在做甜點時須靠大量糖的加入來凸顯它特有的芳香與美味口感。

當法國家庭自家的菜園裏大黃根大量收成，一時間內吃不完，就會製做成家中喜愛的家常果醬。

法國很多省分有着屬於自己地區性的私房大黃根塔，但是，我個人偏愛這種酸甜中夾着烤過酥軟蛋白霜，"阿嬤的老祖宗食譜"的大黃根蛋白霜塔，有着迷人、淺淺的奶油香。五月的下午，走進菜園收割熟成的大黃根，回到廚房，敲敲弄弄同時，法國女伶歌曲伴隨製作甜點。黃昏陽光仍然高照，跟着阿公阿嬤座在庭院的搖椅上，在溫暖且舒服的黃昏陽光照射下享用着來自菜園的美味。

小秘密：阿公説要來杯不甜的白酒最搭這個甜塔了！

Ingredients Fillings (Serves 4)

250 grams Pâté de Sablée
(Crisp Sugar Pastry, refer to p.141)

500 grams Rhubarb

25 grams Butter

50 grams Sugar

Meringue (refer to p154)

125 grams Granulated Sugar

2 each Eggs

Dash Sea Salt

食材內餡 (約4人份)

酥脆甜塔皮 (詳細做法參考 p.141) 250 克、大黃根 (切約 2cm 塊狀) 500 克、奶油 25 克、糖 50 克

蛋白霜 (詳細做法參考法國甜點基本練功夫 6-8)

細砂糖 125 克、蛋白 2 顆、海鹽一小撮

 製作

step 1

將大黃根洗乾淨，瀝乾後切約
2cm 塊狀。鍋子裏裹放進奶油，
加熱融化奶油。

Wash rhubarb, wipe dry and cut into 2 cm pieces. Heat pan and melt butter.

step 2

放進切塊的大黃根翻炒約 10 秒，
加入糖。中小火煮 15 秒，鍋裏
的糖跟奶油變成焦糖即熄火，
備用。

Add rhubarb pieces and stir fry for about 10 seconds. Add sugar and cook under medium heat for 15 seconds. Turn heat off once the butter-sugar mixture had browned. Set aside.

step 3

派皮桿開後，放進塔模裏整好型，塔皮底部用叉子戳洞，放冷藏約 30 分鐘。

Roll out pastry dough. Put into tart mold and fix shape. Prick holes in bottom of pastry with fork. Chill for 30 minutes.

step 4

烤箱以 180℃ 預熱 10 分鐘，整形好冷藏的派皮進烤箱烤 10 分鐘。

Preheat oven to 180℃ for 10 minutes. Put pie dough in and bake for 10 minutes.

step 5

出烤爐，放進剛剛以奶油糖煮過的大黃根，在進烤箱烤 15 分鐘。

Remove pie crust from oven and add buttered rhubarb. Bake again for another 15 minutes.

step 6

蛋白霜製作：一個沙拉鍋裏放進蛋白跟海鹽，將蛋白打成泡末狀後分兩次加入糖用高速打約 25 分鐘左右，直到蛋白霜堅固，紮實不會流動即可。

Meringue : Add egg white and salt into mixing bowl. Beat until foamy. Add sugar, half at a time and continue to beat until stiff peaks formed.

step 7

將蛋白霜放在剛剛烤過的大黃根上，烤箱以 150℃ 烤 25 分鐘。放涼就可以切片享用了！（成品）

Spread meringue over the baked rhubarb. Bake for another 25 minutes at 150℃ . Cool, slice and serve.

Helpful Tips
高亮的小建議

用不完的蛋白霜可以做成脆餅。

Meringue cookies can be made by the unused meringue.

厚皮蛋派

Tarte à gros bords

北部 加萊海岸區
Nord-Pas-de-Calais

> 有一種純粹味道的甜點，不斷地在腦海裏浮現那樣美好滋味。濃厚不造作的口感，透過蛋與乳香味交織，融合在濃郁的厚皮蛋派裏，勾起了我小時候的回憶，奶奶在廚房裏忙碌製作着屬於奶奶與我的獨特味道蛋派……這是我小時候的味道，也是想念我奶奶的味道。

中世紀末期，在法國北方的家庭聚會，尤其是在主教瞻仰巡禮遊行的聚會（在北法的小鄉鎮非常有名的活動），法北的家庭主婦們會製作一些不同口味的厚皮蛋派，或者從甜點麵包店買已經製作好的派或是塔，由於，那時候的家庭沒有錢購買烤模，因為這樣，婦女們將已經做好的派麵皮直接貼在火烤爐的鐵皮旁，為了讓派皮可以堅固，安全無恙的並讓蛋奶液，能夠完整的烤好，就將派皮做的比較厚跟比較高一些，再將蛋液放入塔或派裏在火烤爐裏一起烤熟，因此有"厚皮蛋派 Tarte à gros bords"的名字。主教瞻仰巡禮遊行現今在法北的一些小鄉鎮每年仍然持續舉辦着，而且在法國是非常有名的宗教活動.每戶人家的婆婆媽媽們也會趁此展現自己的厚皮蛋派的手藝，有些非常具創意的媽媽們會在蛋奶液裏加入與往常不同的口味變化，這樣的活動也變成婆婆媽媽們交換自己的"厚皮蛋派"製作食譜心得的好時機。

Ingredients (Serves 8)

1 each Sugar Pastry

500 ml Milk

125 ml Cream

5 each Egg Yolks

125 grams Granulated Sugar

75 grams Corn Flour

Dash Lemon Rind

Dash Icing Sugar
(for decoration)

食材（約 8 人份）

甜塔皮 1 張、牛奶 500ml、
液態鮮奶油 125ml、蛋黃 5 顆、
細砂糖 125 克、玉米粉 75 克、
檸檬皮少許、糖霜（裝飾用）少許

製作

step 1

烤箱以 180℃ 預熱 10 分鐘，將
甜派皮（做法請見 p.141 香酥甜
塔皮）擀開烤好備用。

*Preheat oven to 180℃ for 10 minutes.
Roll out sugar pastry. See French basic
technique for making of the Crispy Sugar
Pantry Pastry on p.141.*

step 2

準備一只深鍋，放進牛奶跟液
態鮮奶油，放在火爐上以小火
加熱。

Warm milk and cream in low heat.

step 3

在另一個沙拉碗裏放進，糖，玉
米粉攪拌均勻加入五顆蛋黃攪拌
均勻。

*Mix sugar, corn flour and egg yolks in
mixing bowl.*

step 4

慢慢分次將加熱的牛奶跟液態鮮奶油加入剛剛蛋跟糖與玉米粉攪混合裏。用打蛋器充分攪拌均勻。

Slowly add heated dairy mixture into egg yolk mixture. Mix with whisk until combined.

step 6

牛奶蛋液放在一旁一邊攪拌讓它有點涼之後，倒入事先已經烤好的塔模裏。

Stir mixture until it is slightly cooled. Pour into baked tart shell.

step 7

放進烤箱烤 25~30 分鐘，放涼後，撒上檸檬皮跟糖霜。

Bake for another 25 to 30 minutes. Cool and then sprinkle lemon rind and icing sugar on top.

step 8

待涼再切，比較可以切出完整的形狀。

Slice when cooled. It cuts up much better that way.

step 5

在將攪拌均勻的牛奶蛋液倒回鍋子裏，放上火爐上，中小火煮，一邊煮一邊攪拌至牛奶蛋液成稠狀即可馬上離火。

Put mixture back into pot and heat with medium heat . Continue to stir with whisk until mixture thickens. Remove from heat.

Helpful Tips
亮亮的小建議

1. 可以在第三步驟時加入一些檸檬皮或是甜橙皮，增加香氣。

2. 第五步驟回到鍋上煮的時候，火爐上的火千萬不能開大火煮，要中小火，若是火太大會焦鍋底。而且手上的打蛋器不能停，要快速的攪拌，因為，牛奶蛋液加熱很快變稠，若是停下來，鍋緣邊邊容易結塊，吃的時候很容易吃到有塊狀的內餡。

3. 如果要內餡有其他的口味，如巧克力或是檸檬，甜橙口味……可以在第三步驟加入即可。

4. 若是沒有玉米粉可以用馬鈴薯粉替代（請勿用太白粉）。

1. *You can add a little lemon or orange rind in step 3 to give it more fragrant.*

2. *On step 5, when pot is put back onto the heat, make sure it is on medium and not high. Otherwise, it will burn the batter. Whisk cannot stop stirring as mixture thickens very quickly. If stirring stops, thick pieces will collect on side of pot and affect the taste of the dessert.*

3. *If you prefer other flavors in the fillings such as chocolate, lemon or orange, you can add them in step 3.*

4. *Corn flour can be replaced with potato starch if not available.*

法國地方性家常特色塔派

芙拉慕斯蘋果塔

芙拉慕斯
蘋果塔
La flamusse

中部 勃艮地 Bourgogne

勃艮地這道家常蘋果塔的法文名稱 La flamusse，也會有人叫它做 La flamous 或是直接稱呼 flamusse。說是甜點，也有些人會做成鹹的口味。水果可以用其他的瓜果類替代，例如：南瓜泥，或是栗子南瓜、大南瓜泥等等。也可以將大南瓜去皮切塊放入來烤，就是美味的勃艮地簡單的午餐，在搭配新鮮翠綠的沙拉就是非常的勃艮地當地家常簡單午餐吃法。

小小午睡片刻後，走入廚房隨手拿起三顆已經久放的青蘋果削皮切塊，奶油入鍋炒過後快速製作出勃艮地美味家常蘋果塔……跟朋友喝茶聊天吃著蘋果塔渡過美好悠閒鄉村的午後時刻。

Ingredients (Serves 3)

3 each Green Apples

2 each Eggs

250 ml Milk

1/2 tbsp Whipping Cream

38 grams Granulated Sugar

8 grams Butter

Dash Sea Salt

25 grams All Purpose Flour

食材（約 3 人份）

青蘋果 3 顆、蛋 2 顆、牛奶 250ml、液態鮮奶油 1/2 湯匙、細砂糖 38 克、奶油 8 克、海鹽一小撮、麵粉 25 克

step 1

蘋果洗淨，去芯後切塊，烤箱以 150℃預熱 10 分鐘。

Wash apples. Cored and cut into pieces. Preheat oven to 150 degrees for 10 minutes.

step 2

用 5 克的奶油將蘋果炒過一下，倒出放在廚房紙巾上吸油後，備用。

Saute apples with 5 grams of butter. Drain on paper towel. Set aside.

step 3

準備一個沙拉碗，放進蛋跟糖、海鹽、牛奶先攪拌均勻後，麵粉過篩加入，最後加入液態鮮奶油。

Add egg, sugar, salt and milk into mixing bowl. Mix and then add sifted flour. Add cream and mix well.

製作

step 4

烤盤裏塗上剩下的 3 克奶油，平均放入用奶油煎過的蘋果。

Coat baking pan with remaining 3 grams of butter. Layer sauteed apples onto pan.

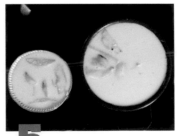

step 5

平均倒入蛋奶液到每個烤模裏。

Pour egg mixture into baking pan.

step 6

送進烤箱烤 45 分鐘，上色後出爐，待涼後撒上糖霜即可享用。

Bake for 45 minutes until the top browned. Cool and sprinkle with icing sugar. Serve.

反烤蘋果塔

Tarte tatin

 中部 Centre

1926 年 一位美食評論家庫爾儂司基在旅行到拉末特-博夫宏時發現了一家餐廳，由兩位未婚的老婦人 Tatin 姊妹共同經營。當他正在品嚐甜點時，端上桌的蘋果塔是他從未享用過的，也未曾聽說過的——焦糖化的蘋果餡在派皮下，香脆餅皮則在蘋果餡料上。友人跟他說這是倒轉過來的蘋果塔。庫爾儂司基發現這塔真是奇蹟般的美味！

一回到巴黎，我們這位喜歡開玩笑的評論家編造初一則故事。他說，因為塔當姊妹中的史蒂芬妮在蘋果塔出爐時可能是一時不小心讓蘋果塔掉落在地上，拾起時又不小心上下顛到了！庫爾儂司基的朋友們就這麼盲目的相信了這個故事並傳遍了全世界。就這樣，塔當蘋果塔就掉入了糕點的神祕傳說裏。

——資料來自 Maguelonne Translator-Samat 的 La très belle et très exquise histoire gâteaux et des friandises 一書

> 昨晚的大風將後院蘋果樹上的蘋果吹掉落一地，今早，提着籃子帶着拉奇走進後面果園拾起被風吹落一地的蘋果，準備下午來製作焦糖化的蘋果塔……喝着濃濃香草味的茶。

Ingredients (Serves 4)

1 each(250 g) Crispy Sugar Pastry

1 kg Apples

125 grams Brown sugar

60 grams Butter

2 tbsp Water

食材 （約 4 人份）

香脆塔皮 1 張約 250 克、
蘋果 1 公斤、紅蔗糖 125 克、
奶油 60 克、水 2 湯匙

製作

烤箱以 180℃ 預熱 10 分鐘。

Preheat oven to 180℃ for 10 minutes.

平底鍋裏放進糖跟兩湯匙水，中
小火慢慢將糖煮成糖漿後放進奶
油，離火讓糖漿稍冷。

Add sugar and 2 tbsp water into frying pan. Cook over medium heat until it turns into a syrup. Add butter. Remove from heat and let it cool.

蘋果洗淨，去皮，去芯，並切成
大塊狀。

Wash apple, remove skin and core. Cut into large pieces.

放進切塊蘋果，回到火爐上煮
15 分鐘。

Add cut apples into syrup. Continue to cook for 15 minutes.

烤盤上放上一張烘焙紙，放上煮
好沾滿糖漿的蘋果跟糖漿汁。

Line baking pan with parchment paper. Pour syrup coated apples into pan.

step 6

將派皮擀平後,蓋在蘋果上方,
將塔邊的派皮整理摺好。

Roll out tart pastry and cover baking pan. Remove excess tart pastry.from pan.

step 7

送進烤箱烤 25 分鐘之後,拿出
烤箱,冷卻後倒扣在盤子上,
切片放上一球香草冰淇淋一起
享用。

Bake for another 25 minutes. Remove from heat. Flip over once it is cooled. Slice and serve with a scoop of vanilla ice cream.

Helpful Tips
亮亮的小建議

1. 若你喜歡甜些的口感的蘋果塔,可以將紅蔗糖更換成細砂糖,細砂糖煮的時候顏色會更深一些,但是紅蔗糖會比較有香氣。

2. 法國人喜歡在反烤蘋果派放上一球香草冰淇淋很適合夏天享用,冬天則可以先將蘋果派放得稍涼一些,再擠上法式香堤也很好吃。

3. 蘋果的選擇,青蘋果會比較酸一些,水份也會比較少。紅蘋果水份比較多,再煮的時候需要花一點時間將糖漿煮收掉一些在倒入烤模裏。如果可以的話選表皮已經老化的蘋果來製作味道也會更美味。

1. *If you prefer a sweeter taste, you can replace brown sugar with granulated sugar.*
   The color will be a little darker but brown sugar will be more fragrant.

2. *For a cooler taste, a scoop of vanilla ice cream is suggested to accompany the dessert on a Summer day. During Winter time, French Crème Chantilly can be added on top once the pie is slightly cooled.*

3. *As for the selection of apples, green apples have a slightly sour taste and less liquid than the red apples. It will take slightly longer to reduce the liquid when red apples are used. If possible, use aged apples which will improve the flavour .*

阿爾薩斯式
蘋果派
Tarte aux pommes
à l'alsacienne

東北部 阿爾薩斯 Alsace

法 國最有名的蘋果甜點除了"反烤蘋果派"之外，還有更多的不同做法的蘋果派。就像法國東北部這道：阿爾薩斯式的傳統蘋果派，有着鮮奶油，雞蛋、牛奶跟糖滑順，細嫩的內餡猶如慕斯與布丁之間的雙重口感，烘烤後微酸的青蘋果口感搭配上酥脆的塔皮是這道甜點不膩口的主要因素。青蘋果與雞蛋奶液內餡的融合，足以展現這道甜點的美妙。

白雪公主只喜歡吃紅蘋果嗎？試試青蘋果，更清香，酸甜……我家拉奇 Lucky 很愛的烤蘋果甜點，一次能吃掉兩份蘋果派。

Ingredients (Serves 6)

250 grams Crispy Sugar Pastry
4~5 Each Medium size apples
125 grams Eggs (about 3)
125 grams Milk
125 grams Whipped Cream
60 grams Granulated Sugar
Dash Vanilla Essence

食材 （約 6 人份）

香脆派皮（甜派皮）250 克、中型蘋果 4~5 顆、雞蛋（約3顆）125 克、牛奶 125 克、霜狀鮮奶油 125 克、細砂糖 60 克、香草精適量

step 1

將派皮擀平,放進塔模裏,用手指將派皮壓貼塔模,塔頂端多餘的派皮可以直接剪掉。叉子在塔皮打布插出許多洞孔,放進冰箱約 20 分鐘後,烤箱 180℃ 預熱 10 分鐘。

Roll sugar pastry. Fit into tart mold. Use fingers to make sure pastry fits tightly into mold. Prick holes into pastry shell with fork. Cool in fridge for 20 minutes. Preheat oven to 180℃ for 10 minutes.

step 3

蘋果塊放進塔皮裏,進烤箱烤 15 分鐘~20 分鐘。

Add apples onto tart shell. Bake for 15 to 20 minutes.

step 5

蛋奶液倒入已經烤過的蘋果跟塔派裏。

Pour the milk egg mixture into the baked tart pastry.

step 6

進烤箱烤 25 分鐘之後,冷卻後撒上糖霜就能夠享用了。

Bake into oven for 25 minutes. Take out and cool down. Dust icing sugar on the tart and serve.

step 4

在一個沙拉碗裏放入蛋跟糖混合加入牛奶跟鮮奶油,最後香草精,混合均勻。

Put egg and sugar together in mixing bowl. Add milk and whipping cream to mix. Finally, stir in vanilla essence well.

step 2

蘋果去皮後,去掉中芯,切塊。

Peel skin and cored apples. Cut into pieces.

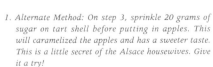

Helpful Tips
亮亮的小建議

1. 有另一種做法:是在步驟三部份的塔皮底層,撒上約 20 克糖再放入蘋果塊進烤箱烤,喜歡甜一些或是喜歡蘋果帶點焦糖香就可以這麼做。這是有些阿爾薩斯的阿嬤們的小祕訣做法,趕快偷偷學起來喔。

2. 如果沒有香草精可以用半根香草莢,對切香草莢取香草籽使用。

3. 另外我自己的小創意做法會將整顆已經削好皮的蘋果放進烤模裏在倒入蛋奶液進烤箱烤,或是不使用塔皮,直接放蘋果跟倒入蛋奶液來烤,也是很不錯的方式。

1. *Alternate Method: On step 3, sprinkle 20 grams of sugar on tart shell before putting in apples. This will caramelized the apples and has a sweeter taste. This is a little secret of the Alsace housewives. Give it a try!*

2. *Vanilla essence can be replaced with 1/2 a vanilla bean. Just scrape the seeds off half the bean.*

3. *My creative way is to use a whole apple, skin removed, put into tart mold and bake in the oven. Then pour egg/milk mixture over. You can also leave out the tart pastry and bake the apple with the egg/milk mixture.*

草莓派
Tarte aux fraises de Plougastel

西部 **不列塔尼 布列斯特 Plougastel**

這個草莓塔最原始來自智利。法國最好品質的草莓則來自不列塔尼的布列斯特小鎮，這裏的氣候適宜，黏土特性的土壤很適合種植出美麗又可口的草莓。因此，這兒的草莓採收可以從春天一直採收道夏天，也因為如此，布列斯特小鎮上的草莓相關甜點都相當美味。

夏天，採草莓的季節，小朋友穿着可愛小雨鞋提着小籃子在草莓園裏那頭聞着草莓，大喊草莓好香喔！草莓園的那頭，拉奇也開心地四處聞聞，順勢找了一個最佳的位置坐下做做日光浴，並朝着我們這邊對着大家笑，似乎也再說草莓園裏的太陽好香，有草莓味……。夏天，是小朋友＆狗狗歡樂的季節，也是草莓的季節，更是小朋友大吃草莓的季節。

{ **Ingredients** (Serves 6)

1 each Crispy Sugar Pastry
130 grams Granulated Sugar
2 each Eggs
750 grams Strawberries
Dash Vanilla Sugar
100 ml Milk
Dash Sea salt

食材（約 6 人份）
香脆甜塔皮 1 張、細砂糖 130 克、蛋 2 顆、草莓 750 克、香草糖少許、牛奶 100ml、海鹽少許

製作

step 1

將塔皮擀開後，放進烤模裏，用手指頭將派皮壓貼在烤模上，塔皮底層用叉子戳洞，放進一張烘焙紙放進足夠的豆子，烤箱180℃預熱10分鐘後，送進塔烤10分鐘，烤出爐後放涼待用。（香脆塔皮詳細做法請見 法國基本練功夫 6-2）

Roll out pastry shell and fit into baking tin. Prick holes inside of pastry shell with fork. . Add parchment paper on top and add enough beans or pie weight into tin. Preheat oven to 180℃ for 10 minutes. Bake shell for 10 minutes. Remove from oven. Set aside and cool. (For detailed method of making tart pastry, refer to French Basic Baking Techniques 6-2)

step 2

在沙拉碗裏放進 2 顆蛋與 80 克的糖、牛奶、香草糖與一小撮海鹽一起打散攪拌和勻。

Mix 2 egg, 80 grams of sugar, milk, vanilla sugar and salt in mixing bowl.

step 3

攪拌好的蛋奶液倒入事先烤好的派塔裏，進烤箱烤 10 分鐘，烤到表面上色就可以出爐了！

Pour egg mixture into baked pie shell. Bake another 10 minutes until browned on top.

step 4

在等待剛烤好的塔放涼的時候，將草莓洗淨，擦乾並將草莓的蒂頭切掉。擺放再剛剛烤好的塔上。

While the pie is cooling, wash the strawberries. Wipe dry and cut off the stem and leaves. Decorate baked pie with strawberries.

step 5

草莓糖漿製作：10 顆草莓用叉子壓碎加上 50 克的糖一起放進一個鍋子裏將糖煮化開即可。趁熱使用刷子沾取草莓糖漿刷在草莓塔上的草莓上，涼了之後就可以享用了！

Strawberry syrup: Mash 10 pieces of strawberries. Add 50 grams of sugar and cook in saucepan until sugar dissolves. Brush syrup on top of pie while it is still hot. Cool and serve.

Helpful Tips 亮亮的小建議

1. 如果沒有香草糖可以用香草粉取代。
2. 為了使草莓塔能夠凝結在塔裏，我們也可以在蛋奶液的製作裏加入 30 克的玉米粉，也會相同的產生凝結的效果。

1. Vanilla powder can be substituted for vanilla sugar.
2. In order to ensure the strawberry can hold together, we can also add 30 grams of corn flour to egg mixture.

藍莓塔

Tarte aux myrtilles

中部 奧弗涅 Auvergne

每年夏天在 Auvergne 地區，從 Haute-Loire 的樹林跟樹林外，包括許多樹林外的小徑生長了許多的藍莓，採收期可以一直持續到七月底或是八月初。每到盛產季，這個可口又多汁的果實實在太多多到讓許多的農夫們傷透腦筋，除了製作冷凍果實來保存，也生產美味的藍莓果醬，也因為盛產過剩的緣故，大家想出製作出美味的藍莓塔，讓美味的藍莓可以多一種享用的方式。

微涼的下午四點，姪女們跟拉奇在後院一陣追逐完後，草地上席地而坐，一手拿着湯匙，另一手端着盤子，盤上微酸甜，新鮮的藍莓塔，加上一球冰淇淋淋上自製淋醬，姪女們吃的津津有味……你一言我一句的訴說如何喜歡冰淇淋跟藍莓在一起的味道。

{ **Ingredients** (Serves 6)

250 grams Crispy Sugar Pastry

500 grams Fresh or frozen blueberries

100 grams Granulated sugar

2 tbsp Icing sugar

125 grams Almond Powder

3 each Egg yolks

2 tbsp Water

{ **Sauce**

150 grams Raspberry or Mulberry Jam

1 tbsp water

1 handful Granulated sugar

食材（約 6 人份）

甜派皮 250 克、新鮮藍莓或是冷凍 500 克、細砂糖 100 克、糖霜 2 湯匙、杏仁粉 125 克、蛋黃 3 顆、水 2 湯匙

淋醬

覆盆子或是桑葚凝醬 150 克、水 1 湯匙、細砂糖一大撮

step 1

藍莓洗乾淨，瀝乾水份。

Wash blueberries. Drain .

step 2

將甜派皮擀開，派皮放上烤盤上，派皮貼緊塔模，去掉多餘的派皮，派底用叉子搓出洞，放進冰箱冷藏，備用。

Roll out pastry dough. Fit into tart tin and remove excess dough. Prick bottom with fork. Refrigerate.

step 3

在一個鋼鍋裏，放進杏仁粉，細砂糖跟蛋黃，用叉子攪拌均勻，加入一點點水再度攪拌。倒入派底，鋪平。

Mix almond powder, sugar and egg yolk in mixing bowl. Add a little water and mix well. Pour into tart tin. Flatten.

step 4

藍莓與糖粉混合後倒入。

Mix blueberries with icing sugar. Pour into tart tin.

step 5

烤箱以 200℃ 預熱 10 分鐘。

Preheat oven to 200℃ for 10 minutes.

step 6

放進烤箱烤 30 分鐘。

Bake for 30 minutes.

step 7

烤的同時間，我們製作淋醬：將凝醬跟水與糖放進小鍋子裏，以小火煮開即可。

While the tart is baking, we will make the sauce: Mix jam with water and sugar. Cook on low heat until combined.

step 8

藍莓塔出爐後淋上淋醬，待冷後撒上糖霜，享用。

Pour sauce over tart when done. Sprinkle with icing sugar when cooled. Serve.

Helpful Tips

亮亮的小建議

1. 不喜歡太甜的朋友，如果覺得淋醬會太甜可以不用製作淋醬，烤好的藍莓塔冷卻後撒上糖霜即可。

2. 淋醬非必須在藍莓塔出來後淋上，你也可以等待藍莓塔涼之後，切片在盤子上，依照個人喜好添加淋醬的份量，也是可以的！

1. If you do not favour the sweet taste, you can omit the sauce. Sprinkle icing sugar on cooled tart.

2. Sauce can also be set on the side for guests to add onto according to one's liking.

西南部 阿爾薩斯洛林 Alsace-Lorraine

白乳酪塔
Tarte au fromage blanc

白乳酪塔 Tarte au fromage blanc 在阿爾薩斯省是一道非常著名的甜塔糕點。許多阿爾薩斯人會在這道白乳酪塔裏加入少許的食用鹽或是少量的櫻桃酒，部分的人製作時會加入葡萄乾或是乾煎熟的杏仁果，每位阿爾薩斯奶奶們都會有自己的私房做法，這真的是一道簡單又美味的法國的家常派。

Ingredients (Serves 6-8)

250 grams Savory Pie Crust

150 grams Granulated sugar

4 each Eggs

50 grams All Purpose Flour

100 grams Melted Butter

500 grams White Yoghurt

1/2 each Lemon Rind

食材（約 6~8 人份）

鹹派皮 250 克、細砂糖 150 克、
蛋 4 顆、麵粉 50 克、融化奶油 100 克、
白乳酪 500 克、檸檬碎皮 1/2 顆

 製作

step 1

將派皮開約 5mm 厚度，烤模上
塗上奶油。

*Roll tart crust to 5mm thickness. Brush
baking tin with butter.*

step 2

將派皮放上烤模上派皮貼緊烤
模。派皮底用叉子戳洞，放進冰
箱冷藏。

*Fit tart crust onto baking tin. Prick
holes on bottom of crust with fork.
Refrigerate crust.*

step 3

蛋黃跟蛋白分開放，一個沙拉碗
裏放進白乳酪、糖、蛋黃、麵粉
跟融化奶油攪拌均勻。

*Separate egg white and yolk. Add
yoghurt, sugar, egg yolk, flour and butter
into mixing bowl. Combine.*

step 4

加入檸檬碎皮，再度攪拌均勻。

Add lemon rind. Mix until combined.

step 5

另一只鍋裏放進四顆蛋白在多加1湯匙的蛋白與加入少許鹽電動打蛋器將蛋白打發，打到電動打蛋器的攪拌棒將蛋白拉起來不會往下掉即可。

In a separate bowl, add 4 egg white plus an additional tablespoon of eggwhite and a dash of salt. Beat until stiff peaks formed.

step 6

慢慢加入剛剛攪拌混合的白乳酪麵糊裏，一邊加一邊攪拌，請務必輕輕攪拌。

Gently fold beaten egg white into yoghurt mixture.

step 7

烤箱以180℃預熱10分鐘，派皮先進去烤15分鐘，出爐後加入白乳酪麵糊，再度進烤箱，以175℃烤約30分鐘。

Preheat oven to 180℃ for 10 minutes. Bake tart crust for 15 minutes. . After it is done, add yoghurt batter onto crust. Bake at 150℃ for 30 minutes.

step 8

出爐後，等冷卻後撒上糖霜，就可以享用。

Let tart cool and then sprinkle with icing sugar. Enjoy.

亮亮的小建議

1. 白乳酪打開的時候，如果裏面有點白乳酪的水，將水瀝掉，不要加入麵糊裏一起攪拌。
2. 法式白乳酪吃起來比美式乳酪的口感還要輕盈，喜歡更多層次口感的話，建議淋上莓果果醬一起享用，口味會更豐富。
3. 烤白乳酪塔的時候，白乳酪會膨脹脹高，若是離烤箱的熱管太近有可能會讓白乳酪爆裂，烤的時候如果上色太快，可以蓋上一張烘焙紙在塔上面繼續烤，有幫助白乳酪減緩上色。或是將烤架在往下降一層。白乳酪塔出爐後，膨的狀況會慢慢消退。
4. 如果你的烤模比較大或是比較高，烤的時間需要增加10~15分鐘來烤熟蛋糕喔啊！
5. 千萬不要將剛出爐時熱熱的將蛋糕切片，乳酪蛋糕需要一個晚上的時間冷卻，隔天享用風味會更好。

1. *If you see liquid in the yoghurt when you open it, drain the liquid first. Do not add into the batter.*
2. *French yoghurt is lighter than American yoghurt. If you want to improve the texture of the yoghurt, a suggestion is to add strawberry jam to it.*
3. *When you are baking the tart, the yoghurt batter will rise pretty high. If it is too close to the heating element in the oven, the batter may crack. If it browns too quickly, cover the top of the tart with a piece of parchment paper. This will slow down the browning. Or you can lower the cooking rack. Once the tart is down, the expansion will slow down.*
4. *If your tart tin is slightly bigger or higher, you will need to increase cooking time for 10-15 minutes*
5. *Cool tart overnight before cutting it. It will taste much better the next day.*

焦糖核
果仁小塔
Tartelettes aux
mendiants

 普羅旺斯 蔚藍海岸
Provence Cote d'Azur

焦
糖核果仁塔是南法地區歡慶聖誕節時享用甜點裏的其中一樣。在塔底的位置主要放乾果子的。傳統的南法聖誕節裏甜點總共有 13 道甜點，其中幾項甜點必須使用四種不同顏色的乾果：杏仁、夏威夷豆、無花果跟葡萄乾，因為這四種乾果的顏色分別代表着在中世紀時期四個不同教會修道士的修行服顏色，雖然現在大家對這四種修道士的修行服顏色記憶存在有點模糊的記憶，但是；南法許多老一輩的長輩們仍然還是遵循傳統，這四種地區教會：Carmes，Augustins，Franciscains 與 Dominicains。

	代表色	辨識果
Franciscains	灰色，近乎鐵灰色	葡萄乾
Dominicains	潔白色	杏仁果
Augustins	紫色	無花果或是其他紫色的乾果也可以使用
Carmes	咖啡色	只要是咖啡色果實都可以用

跟着時代的改變，傳統的法國甜點也以不同風貌來呈現詮釋傳統的風味甜點。因此；塔裏果實大家也都依照當地的風俗來做更換與修改。只要是口味上不會偏離太遠，傳統的禮儀還是存在的。

> 下方的食譜是我家阿嬤的做法，焦糖風味濃郁，乾果味道很豐富。我十分推薦是我跟其中之一最愛的法國小塔甜點其中之一。
> 一本好書，一杯好的咖啡跟幾個小核果仁塔，在陽光普照的庭院渡過一個美好的下午時光……。

法國地方性家常特色塔派

55

Helpful Tips
亮亮的小建議

在第二步驟時加一小撮的海
鹽，增加焦糖的風味。

*Add a dash of sea salt in step 2. This will
improve the caramel flavor.*

{ **Ingredients** (Serves 4)

125 grams Crispy Sugar Pastry (sweet)

50 grams Butter

25 grams Granulated Sugar

2 Tbsp Honey

25 grams Almond

25 grams Pistachio

40 grams Raisins

30 grams Dried Figs

食材（約 4 人份）

香脆塔皮（甜）125 克、奶油 50 克、細砂糖 25 克、蜂蜜 2 湯匙、杏仁果 25 克、
開心果 25 克、葡萄乾 40 克、無花果乾 30 克

製作

step 1

先將塔皮擀開，放上塔模上，整
出塔模的形狀，並將塔皮壓緊塔
模，用叉子在塔皮底部插出氣
孔。放進冰箱冷藏，待用。

*Roll out tart pastry. Cut pastry with
tart mold. Fit pastry into mold. Prick
bottom of crust with fork.*

step 2

奶油放進一個鍋子裏小火融化奶
油。奶油溶化後加入糖，蜂蜜。

*Melt butter in small saucepan. Add
sugar and honey.*

step 5

用湯匙舀用糖漿煮過的乾果放進塔模裏。

Spoon syrup dried fruits and nuts into tart crust.

step 6

烤箱裏烤約 20 分鐘。出烤箱冷卻後就可以搭配咖啡或是茶一起享用囉！

Bake for another 20 minutes. Once they are cooled, enjoy them with coffee or tea.

step 3

無花果乾切小塊跟其他的乾果一起放進鍋子，跟着糖漿煮一會兒，可以用木匙攪拌。

Dice dried figs, Add all dried fruits and nuts into saucepan. Cook and stir slightly with syrup.

step 4

烤箱以 180℃ 預熱 10 分鐘，拿出冰箱的塔派。

Preheat oven to 180℃ for 10 minutes. Refrigerate baked crust.

part 3
法國家常經典
小甜點

法國
鮮奶油泡芙
Choux à la crème Chantilly

北部 皮卡地 Picardie

最早、最傳統的圓形泡芙裏夾層內就是使用法式香堤。

法式香堤的發明者是方斯華瓦德肋（François Vatel），他用生奶油跟一種植物的香草味（也就是現在我們說的香草莢／雲呢拿豆）製作出來甜奶油而聞名。在 1671 年的某天裏，香堤城堡的王子路易二世在自己的城堡舉辦宴會招待路易十四與 2000 多人，方斯華瓦德肋是當時城堡的廚師，為了這個奢侈的宴會而發明的甜奶油，也因此成名。甜奶油也就以 Chantilly 城堡的名字來命名為「Crème Chantilly」。這位知名的大廚也曾經服務於路易十四的財政大城尼可拉福克，在服務於香堤城堡之前，方斯華瓦德肋早就成為一位相當具有知名度的大廚。

全家大小，包括 Lucky 都愛的法式香堤 Crème Chantilly，香甜的奶油還有淡淡香草籽優雅迷人的香氣。

Ingredients

(makes about 8 medium sized puffs)

35 grams All Purpose Flour

1 each Egg

50 ml Water

1/2 Tsp Sugar

20 grams Butter

Dash Sea Salt

French Chantilly Cream

200 ml Whipping Cream

30 grams Icing sugar or granulated sugar

3/4 stick Vanilla Bean

食材（約 8 個中型泡芙）

麵粉 35 克、蛋 1 顆、水 50ml、糖 1/2 茶匙、奶油 20 克、海鹽少許

法式香草香堤（鮮奶油）

液態鮮奶油 200ml、糖霜或是細砂糖 30 克、香草莢 3/4 根

 製作

step 1

將水，奶油，糖跟海鹽放進鍋裏
煮滾。

*Add water, butter, sugar and sea salt to
saucepan. Bring to a boil.*

step 2

放進所有的麵粉，用木匙快速攪
拌，將麵糊全部攪拌吸收奶油水
並且成麵團狀。

*Add flour and quickly stir with spatula
until all the flour had been absorbed by
the liquid. and dough formed.*

step 3

離開火源，加入蛋，使用手動打
蛋器把麵糊與蛋打混合一起。

*Remove from heat and add egg. Combine
with handheld mixer.*

step 4

烤箱以 180℃ 預熱 10 分鐘。

Preheat oven to 180℃ for 10 minutes.

step 6

進烤箱烤 20 分鐘，上色就可以出烤箱了。放涼備用。

Bake for 20 minutes. Once browned, they can be removed from oven. Set aside to cool down.

step 7

一個鋼鍋裏放進鮮奶油跟糖與香草籽（香草莢剖半，用刀尖刮出香草籽放入鍋內）電動打蛋器打發（請 p.150 6-5 做法）打發的香堤裝進擠花袋裏。

Add cream, sugar and vanilla bean into mixing bowl. (Slice vanilla bean in half and scrape seeds into bowl). Beat with electric mixer (refer to p.150, 6-5). Fill piping bag with Chantilly cream.

step 5

將麵糊裝進擠花袋裏，烤盤上塗上少許奶油，麵糊擠在烤盤上，要有一定的間距，不然烤的時候，泡芙會全黏在一起。

Fit batter into piping bag. Lightly brush baking pan with butter. Pipe batter onto pan,. Make sure you leave a good distance between each puff . If they are too close, they will stick together when they rise during baking.

Helpful Tips
亮亮的小建議

做好的法國鮮奶油泡芙，可以淋上熱巧克力醬冷熱吃法還蠻受大家歡迎的或是涼涼的英式香草淋醬淋上冰冰的吃。喜歡冰淇淋的話也可以以冰淇淋取代法式香堤，夏天的時候，在法國餐館，餐後我都會非常渴望來一份內層英一球有冰淇淋，再淋上巧克力淋醬的泡芙。

Pouring hot chocolate fudge over the Chantilly puffs is a popular way of serving them. Cool vanilla sauce over cold puffs is an alternate serving suggestion. If you like ice cream, you can replace Chantilly cream with ice cream. During the Summer time, most people will want to order an ice cream puff after dinner at the French restaurants.

step 8

泡芙橫切成兩半，將法式香堤擠在泡芙底層，接着蓋上上蓋，將所有的泡芙擠完就可以了。撒上糖粉就可以享用了！

Slice puff in half horizontally. Squeeze chantilly cream onto bottom half of puff. Cover with top half. Finish all the puffs and sprinkle icing sugar on top. Enjoy!

黑李蛋糕
Gâteau aux
prunes

北部 加萊海峽區 Pas-de-Calais

七月初夏，正剛好是黑李收成季，下午大家忙碌得爬上黑李樹上，一邊採李一邊吃。拿着竹籃裏的黑李來做奶油蛋香氣濃郁的黑李蛋糕。

七月的法北，正值植採收甜又多汁的黑李季節。黑李樹跟蘋果樹、橄欖樹一樣，都種植在每個家庭得前庭院，水果採收季時家家戶戶搬着梯子，往樹上爬，樹下的人就忙碌着搬運採收的黑李。黑李子產量很多的時候，每戶人家的做法會先拿一部份製作私房果醬，阿公有一位朋友杰哈。

去年，我們一起在阿公家庭院採收黑李，我曾向他詢問他家是如何製作黑李果醬。他怎麼樣都不肯正面回答我。幾天後，他來我家，並詢問我黑李果醬做了嗎？問我怎麼做的？我很大方的跟他分享做法。看來，每戶人家真的很保護自家的私房祕方食譜。

黑李蛋糕也是，既然黑李盛產，黑李蛋糕就是家庭裏的首選製作的甜點。法北由於天氣較為冷，因而北邊人不論從料理到甜點，都會充分的運用奶油，來增加身體的熱量。長久下來，奶油吃的量大過於南法跟中部。也就這樣成為法北的許多料理＆甜點上的一大特色。

Ingredients (Serves 6)

6 each (approx 500 g) Black Plums
125 grams All Purpose Flour
90 grams Granulated Sugar
45 Ml Milk

45 ml Olive Oil
3 each Eggs
1 tsp Aluminum Free Baking Powder
80 grams Butter

食材（約 6 人份）

黑李 6 顆約 500 克、麵粉 125 克、細砂糖 90 克、牛奶 45ml、
橄欖油 45ml、蛋 3 顆、無鋁泡打粉 1 茶匙、奶油 80 克

製作

step 1

將黑李對切，去籽。

Cut plums in half and remove seeds.

step 2

在一個沙拉碗裏放進麵粉、60 克
糖、泡打粉、牛奶、橄欖油與兩
顆蛋攪拌均勻。

*Add flour, 60 grams sugar, baking
powder, milk, olive oil and two eggs.
Mix thoroughly.*

step 3

烤模裏放入黑李，請將黑李皮那
面朝上，果肉朝烤模放入。

*Pour plums into baking mold, skin
facing up and meat facing down.*

step 4

到入剛剛拌好的麵糊，烤箱以
180℃ 預熱 10 分鐘烤 20 分鐘。
（請放在烤箱第四層烤）

*Pour batter into mold. Preheat oven
to 180℃ for 10 minutes. Bake for 20
minutes. (Please put baking rack on the
fourth level)*

step 5

趁着黑李進烤箱，這時我們將奶
油融化，融化後的奶油加入糖
跟打散混合的蛋液，攪拌均勻。
備用。

*While cake is baking, melt butter. Add
remaining sugar and beaten eggs. Mix
and set aside .*

step 6

黑李出烤箱後，淋上奶油蛋黃糖
液，在進烤箱烤 15 分鐘。

*When cake is done, remove from oven
and pour butter syrup on top. Bake for
another 15 minutes.*

step 7

出烤箱後，放涼就可以切片享
用了。

Remove from heat. Cool, slice and serve.

法國家常經典小甜點

熔岩巧克力
Fondent au chocolat

巴黎

> 沒有任何一件事，比得上享受法式"熔岩巧克力"還要來得令人感到開心……。

熔岩巧克力在很早以前就已經出現在法國家庭餐桌上。當年，它是非常不受喜愛甜點的法國人青睞與喜愛。然而；在一位法國的 chef 宴客機會下，才讓這道猶如灰姑娘般的熔岩巧克力變成如今有名氣。

一開始的誕生就不被大眾接受的"熔岩巧克力"，這道不受大家喜歡的甜點在 Chef Michel Bras 精心研究改良，約有兩年的時間下完美地呈現在大家面前。最早做法是用餅乾碎來製作外層的派皮，裏面則是用甘那許巧克力醬。直到 1981 年，也正是熔岩巧克力新生的時候。那時的熔岩巧克力已經被修改為容易且非常簡單製作了，最主要的構成是由兩個餅乾派皮，再使用兩種不同的烤溫。演變到現在則用一種餅乾派皮與非常濃郁的巧克力醬內餡。當然還有更簡單，更完美的做法關鍵就是在於"溫度"，熔岩巧克力千萬不能使用恒溫來烤它，要用高溫來製作，才會流出猶如岩漿般的巧克力醬。

如果你是第一次製作，而且失敗了，千萬別灰心要堅持繼續下去，你將會得到美好的結果，這個結果會讓替你帶來滿心的喜悅。

Ingredients (Serves 4)

100 grams Chocolate

100 grams All Purpose Flour

80 grams Butter

50 grams Granulated Sugar

2 Each Whole eggs + 2 Each egg Yolks

Dash Cocoa Powder

食材（約 4 人份）

巧克力 100 克、麵粉 100 克、
奶油 80 克、細砂糖 50 克、
全蛋 2 顆＋蛋黃 2 顆、可可粉少許

製作

step 1

巧克力敲碎，放進一個沙拉碗裏
跟奶油一起；另外用一個比沙拉
碗還要大的鍋子，放入少許水，
將放有奶油跟巧克力的沙拉碗放
入，一起放在火爐上做"隔水加
熱融化"。

*Chop chocolate into pieces. Add butter
to chocolate in a stainless steel mixing
bowl. Use another slightly larger mixing
bowl and add water to it. Put chocolate
and butter bowl onto the larger bowl.
Melt over heat using the water bath
method.*

step 2

另一個沙拉碗放入蛋跟糖，使用
手動打蛋器打到蛋糖混合呈淺黃
色，等待備用。

*Add eggs and sugar to another mixing
bowl. Beat with mixer until mixture is
pale yellow. Set aside.*

step 3

將融化的奶油跟巧克力的沙拉碗
從大鍋取出，稍放涼。接着，蛋
糖液倒入融化的奶油跟巧克力鍋
裏，請務必一邊攪拌一邊倒。

*Remove melted butter and chocolate
from the large bowl. Let cool slightly.
Add egg sugar mixture into melted
chocolate. Keep stirring while mixing
them.*

step 4

之後，麵粉過篩後倒入並且攪拌混合均勻，要完全將麵粉拌入巧克力糊裏。

Sift flour and mix into chocolate mixture. Mix thoroughly.

step 5

烤模底層＆周圍都塗上奶油，並撒上少許的可可粉方便等一下小蛋糕好脫膜。倒入拌好的麵糊約8分滿即可。

Brush the bottom and sides of baking mold with butter. Also sprinkle some cocoa powder for easy release. Pour batter onto mold , about 80% full.

step 6

烤箱以 200℃ 預熱 10 分鐘後，將巧克力麵糊送進烤箱烤 10 分鐘。

Preheat oven to 00℃ for 10 minutes. Bake batter for 10 minutes.

step 7

出爐後，用刀尖小心的在烤模邊緣劃開，再倒扣仕盤子上。

Once it is done, use a sharp knife and slowly run it around the side of the mold. Flip over onto the serving plate.

step 8

享用時，建議你淋上英式香草醬或是一球香草冰淇淋在幾顆核桃一起享用。

A serving suggestion is to pour English Vanilla Sauce or a scoop of Vanilla Ice Cream with several pieces of walnuts.

Helpful Tips　亮亮的小建議

1. 巧克力的選擇，最好使用可可成份 70% 以上的來製作會更好。

2. 在步驟 3 融化奶油跟巧克力的沙拉碗放涼的程度測試，可用你的手觸摸沙拉碗底，如果你的手可以承受碗底的熱度，表示已經可以將蛋糖加入。

3. 完成後的熔岩巧克力，如果你沒有事先準備英式香草醬或是冰淇淋，也可撒上糖粉或是可可粉，讓成品更有完整性與美感，會與餐廳裏的甜點很像喔。

1. It is best to select Chocolate with over 70% of Cocoa .

2. On step 3, chocolate mixture should be cool enough so your hand can tolerate the heat on the bottom of the pan. It is safe to add the egg sugar mixture in.

3. Instead of English Vanilla Sauce or ice cream, you can also sprinkle fondent with icing sugar or cocoa powder. This will improve the presentation of the dessert .

倒扣焦糖布丁
Crème renversée au caramel

巴黎

倒扣焦糖布丁，也有人稱作"雞蛋布丁"或是"焦糖雞蛋布丁"。 主要都是以蛋、糖、牛奶跟香草莢混合烤出的滑順的內餡體。牛奶蛋糖餡裏沒有太多的糖分，若是不喜歡吃太甜的朋友可以在這部份在做縮減糖份，只要保留焦糖的甜度就好。享用時，最好是用湯匙一口氣舀起焦糖跟布丁體，焦糖的甜度會與不太甜的布丁體混合達到食用的甜味平衡。製作這道甜點通常會使用一般白色細砂糖。細砂糖比黃砂糖甜度高；但是，黃砂糖卻比較有蔗糖香氣，如果喜歡焦香氣味重一點的話，不妨可以將細砂糖更換成黃砂糖喔！

這是一道法國家家戶戶都會做的甜點，老人家們的食譜也有許多不同之處，每份食譜做法得最終成果都吃得出來每位奶奶們對家中孫子、小孩滿滿的愛。法國奶奶們喜歡將倒扣焦糖布丁裝放在大的烤模裏，烤好時，也不倒扣了！大家圍着餐桌拿着湯匙挖搶着吃……是不是很有家庭歡樂的氣氛呢！

下午，跟着阿嬤走到母雞下蛋的專屬小房子裏，撿在手上溫溫的，沾滿稻草與糞土外殼的雞蛋，趁着雞蛋新鮮，阿嬤想着如何運用這些新鮮的蛋做甜點給等等回家的小孫子吃……。

Ingredients (Serves 4)

3 each Whole eggs + 1 egg yolk

500 ml Whole Milk

1/2 each Vanilla Bean

100 grams Granulated sugar

Caramel

100 grams Granulated Sugar

1 tbsp Water

食材（約 4 人份）

蛋 3 顆，外加 1 顆蛋黃、
全脂牛奶 500ml、香草莢 1/2 根、
細砂糖 100 克

焦糖

細砂糖 100 克、水 1 湯匙

製作

step 1

準備一只深鍋放進糖跟水，小
火慢慢將糖煮化，轉成焦糖色
後離火，倒入烤模裏，放涼，
等待使用。

Add water and sugar to saucepan. Cook over low heat until sugar caramelized. Remove from heat. Pour into baking mold. Cooled and set aside.

step 2

另一只鍋子裏，放進牛奶，香
草莢對切後用刀尖將香草籽刮
出來放進牛奶鍋裏，香草莢的
外皮一同放入牛奶鍋內，小火
煮滾，煮到鍋子邊邊起泡泡就
可以離火了。

In another saucepan, add milk, vanilla seeds scraped from vanilla bean and vanilla bean skin. Cook over low heat and bring to a boil. When the liquid on the side of the pan starts to bubble, it can be removed from heat.

step 3

沙拉碗裏放入雞蛋跟糖，使用
手動打蛋器將蛋跟糖打混合在
一起。

Beat egg and sugar with hand held beater in mixing bowl until combined.

step 6

烤箱以 180℃ 預熱 10 分鐘。將蛋奶液倒入事先有倒入焦糖的烤模裏。蛋奶液只要倒入 8 分滿即可。

Preheat oven to 180℃ for 10 minutes. Pour egg mixture into mold with caramel. Liquid should only be 80 % full.

step 4

將煮滾的香草牛奶倒入蛋糖裏攪拌均勻。

Pour heated vanilla milk into egg sugar mixture. Combine.

step 5

混合的蛋奶液，用過篩的篩子過濾後取出香草莢外皮，放涼一些，等待倒入烤模裏。

Strain mixture over sieve. Cool for a while and then pour into baking mold.

step 7

準備另一個大烤模裝入到達烤模一半位置的溫水，放入烤模後進烤箱烤 30 分鐘。烤出後放涼，用刀尖的部份沿着烤膜邊緣劃一圈再倒扣在盤子上，用自己喜歡的水果稍加裝飾一下即可。

Prepare a larger baking pan and fill it halfway with water. Put baking mold in and bake for 30 minutes. Cool after it is done. Use a sharp knife and run it around the edge of the baking mold. Flip onto serving plate. Decorate with your favorite fruits.

焦糖
烤蘋果
Pommes
caramélisées

南部　羅納 - 阿爾卑斯山區
Rhone-Alpes

下雨天，空氣裏散發淡淡肉桂香，優雅的蘭姆酒，熱熱焦糖跟蘋果果肉混合在一起入口……下雨天，真適合這個香氣；有着溫暖，短暫幸福的味道。非常簡單，家常卻是非常非常好吃的烤蘋果甜點，這道甜品要好吃最好選擇加拿大品種的斑皮蘋果，也就是蘋果外皮有點黃色斑點或是斑紋，外觀上我們可以看到很多咖啡色跟灰色斑點在蘋果外皮上，選擇這樣的蘋果最佳，它有足夠的蘋果香氣，果肉紮實。紮實果肉的蘋果最適合與焦糖做結合，也能展現出這道甜點的小小層次風味。

Ingredients (Serves 2)

2 each Apples

15 grams Butter

25 grams Brown Sugar

1/2 tsp Cinnamon Powder

Caramel

50 grams Sugar

1 tbsp Water

23 grams Butter

1 tsp Rum

食材（約 2 人份）
蘋果 2 顆、奶油 15 克、紅砂糖 25 克、肉桂粉 1/2 茶匙

焦糖
糖 50 克、水 1 湯匙、奶油 23 克、蘭姆酒 1 茶匙

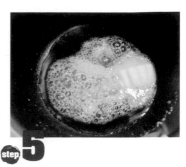

step 1

蘋果洗淨，擦乾後，以橫面平切
方式切掉蘋果蓋頭，使用小湯匙
將蘋果梗＆芯挖掉（請小心要保
持蘋果外觀完整）。

*Wash apples and wipe dry. Cut off top
of apples horizontally. Scoop out core
and seeds with soup spoon. (Be careful
to keep the apples intact)*

step 3

撒上肉桂粉在蘋果上，烤箱以
150℃ 預熱 10 分鐘，烤個 25
分鐘。

*Sprinkle cinnamon on top of apples.
Preheat oven to 150℃ for 10 minutes.
Bake for 25 minutes.*

step 5

在糖水轉變成焦糖褐色時，加入
23 克的奶油跟蘭姆酒。

*When syrup starts to turn caramel
brown, add 23 grams of butter and rum*

step 4

製作焦糖漿：糖跟水放進鍋子
裏，放上火爐加熱。

*Caramel Syrup: Add sugar and water to
saucepan. Heat mixture*

step 2

奶油＆糖放進挖空中芯的蘋果
裏，蓋上蘋果蓋頭。

*Fill cavities of apples with butter and
sugar. Cover with apple tops.*

step 6

將焦糖淋在烤箱裏的蘋果上，在
進烤箱烤約 2~3 分鐘，讓蘋果
上點色。

*Pour caramel syrup over apples still in
oven. Let the apples brown.*

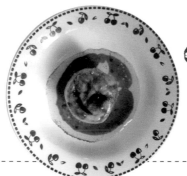

step 7

可以趁熱享用，或是涼涼的食用。

Serve while it is hot or cold.

Helpful Tips
亮亮的小建議

1. 蘋果中心也可以放一些葡萄乾或是核桃一起烤。

2. 蘭姆酒可以替換成白蘭地，如果你喜歡酒香味濃點，可以在最
   後出烤爐的時候，趁熱淋上白蘭地酒，淋酒的烤蘋果很適合冬
   天熱熱享用。夏天，喜歡吃涼的，蘋果烤出爐後等蘋果涼放上
   一球香草冰淇淋或是一球法式香堤 Crème Chantilly 也很受小
   朋友們的歡迎喔。

1. *Raisins or walnuts can also be added into the apples before
   baking.*

2. *Rum can be replaced with brandy. If you want more
   liqueur taste, you can add brandy on top of apples right
   out of the oven. Liqueur over apples is very suitable for
   Winter serving. During the Summer months, a good serving
   suggestion is to let the apple cool and then add a scoop of
   Vanilla ice cream or Chantilly cream .*

甜白酒
蜂蜜烤梨
Poires au miel

南部 普羅旺斯 蔚藍海岸
Provence Cote d'Azur

晚餐後……自後院採下的新鮮洋梨做成美味的餐後甜點。到自己果園採收當天的餐桌食材，是法國鄉村生活裏的一個小小幸福。

如　你有機會找到一種專門製作果醬的西洋梨品種「秋梨」，法文為Martin-sec，小小顆的秋梨，表皮有着特殊的香氣，極類似台灣砂梨或是三灣梨（一般稱為大梨或是粗梨）的外皮顏色。法國Martin-sec品種的秋梨的產地大多在法國與瑞士連接的山區。這種梨子大小一個人幾乎應該可以吃掉兩顆，又因為她的果皮有着香氣，皮稍厚些，烤過之後的梨果肉會有美妙融化的口感。

Ingredients (Serves 4)

4 Each Small Pears

75 grams Raisins

75 grams Almond

3 Tbsp Honey

1 Cup Sweet White Wine

30 grams Butter

食材（約4人份）

小梨4顆、葡萄乾75克、
杏仁果75克、蜂蜜3湯匙、
甜白酒1杯、奶油30克

製作

step 1

烤箱以 170℃ 預熱 10 分鐘。

Preheat oven to 170℃ for 10 minutes.

step 2

洗乾淨梨子表皮並擦乾，去掉梨外皮，並切開梨帽。（梨子帶梗的頂端部位）

Wash and dry pears. Peel skin and cut pear top off horizontally. (Top close to stem)

step 3

用湯匙將梨子裏面的核籽帶芯一並挖掉，千萬不要挖破也不要挖見底喔！

Scoop core and seeds out with soup spoon. Do not break the bottom of the pear.

step 4

在一個小碗裏放入葡萄乾、杏果並加入蜂蜜攪拌混合。

Mix raisins, almonds and honey in a small bowl.

step 5

將混合好的葡萄乾與杏果填入梨子內，蓋上梨帽。放上一小顆奶油在梨帽上。

Stuff mixed fruits and nuts into pears. Cover with pear lids. Add a dollop of butter on the pear lid.

step 6

烤盤抹上奶油倒入，甜白酒，放入填好乾果仁的梨子，進烤箱烤約 20 分鐘。

Brush baking pan with butter. Add wine. Place stuffed pears onto pan. Bake for 20 minutes.

step 7

在烤的過程中，偶爾打開烤箱將烤盤底的白酒舀起來淋在梨子上，讓梨子保持濕潤並有酒香味。

While pears were baking, occasionally open the door of the oven, scoop white wine from pan and pour over the pears. This will allow the pears to remain moist and retain a nice liqueur fragrant.

step 8

出烤箱，盛盤後，淋上些許烤盤內的白酒一起享用。

When done, remove from oven and set onto serving plates. Pour additional wine from baking pan over pears to serve.

南部

普羅旺斯 蔚藍海岸
Provence Cote d'Azur

肉桂
甜橙果泥
Compote d'orange,
cannelle

夏 天南法陽光的滋味,是屬於甜甜的,強眼的陽光與可口的果實……。 我有一個關於"果泥"的故事,好像是在某本法文書中讀到關於來自印度文學裏有這麼一段蘋果果泥的一段小詞句。

"某天早上,我躺在其中一部分的灌木叢堆裏,我喜歡露出上身並將它曬成均勻的顏色,突然發現有一堆蘋果隱藏在這堆草叢裏,都被我壓碎了,我的下巴還輕微的擦傷"。

在中世紀時期,由義大利的阿梅迪奧八世寫下的食譜在書裏的名字:Ung emplumeus de pomes 是煮熟的蘋果泥,這道食譜的傳統做法到現在還被保留在《Du fait de cuysine, 寫於 1420.》一書裏。

> 歐洲製作果泥或是果糊的做法幾乎很接近,都主要以水跟糖煮滾加入水果果肉跟果汁。
> "果泥"就將水果切稍大塊,煮到還可以看到果塊還有一些果泥的狀態。"果糊"就是幾乎將水果果肉煮到變成泥狀,這個在很多家庭會用來餵養小嬰兒,當成小朋友的副食品享用,但要給小朋友食用,則要斟酌糖的份量。

Ingredients (Serves 2)

800 grams Sweet Orange

250 grams Granulated Sugar

2 Each Cinnamon Sticks

食材（約 2 人份）

甜橙 800 克、細砂糖 250 克、肉桂棒 2 根

製作

step 3

一個深鍋放入 250 克的糖跟 100 克水先煮滾後放甜橙果肉，跟肉桂棒與碗裏剩下的果汁。

Bring 250 grams of sugar and 100 grams of water to a boil. Add the orange segments, cinnamon stick and any orange juice remaining in the bowl.

step 1

甜橙洗淨，擦乾後放置切菜板上，先切掉甜橙頭，接着將甜橙皮包含白膜的部份一起切掉，多少會切到一些果肉，沒有什麼關係的。

Wash and dry orange. Cut off the top of the orange. Peel off orange skin and white membrane. It is Ok if some of the flesh is taken out with the skin.

step 2

再將甜橙的果肉與果肉之間的白色薄膜偏往果肉的部份切下一瓣瓣的甜橙果肉，放在一個碗裏。

Peel off all white membrane covering the orange segments. Put all orange segments into bowl.

step 4

中小火煮約 30 分鐘即完成，涼涼的享用，很美味的喔。

Cook over low/ medium heat for 30 minutes. Serve after it is cooled. Very tasty.

Helpful Tips 亮亮的小建議

1. 肉桂棒的部份可以更改成肉桂粉或是八角，程序做法跟肉桂棒相同。

2. 你也可以不放肉桂或是八角，完成後放上新鮮的蒔蘿葉（茴香）切碎撒上，黃黃綠綠的色系很有南法風味。口感也很特別喔！

3. 若是要給小嬰兒當副食品享用，只要水滾後放入甜橙下去煮即可，不用放細砂糖跟肉桂棒。

4. 煎雞肉，豬肉排時，可以拿甜橙果泥來做醃清肉品的醬料，或是做成肉排的淋醬，甜橙味跟肉排都很搭稱，也十分的解油膩。

1. Cinnamon sticks can be replaced with cinnamon powder or anise. Use them like the cinnamon stick.

2. You can omit the cinnamon and anise completely. Once it is done, sprinkle fresh chopped dill on top. Yellow and green, full of taste of Southern France.

3. If you want to serve this to infants, add orange segments to boiling water. Cinnamon and sugar are not necessary.

4. You can also use this to marinate chicken steak and pork chop. It can also be used as a sauce over pork chops. The sweet orange taste matches very well with the pork and also makes it less greasy.

是法國東南部一個大區，南鄰西班牙與地中海，這道加泰隆尼亞烤焦糖奶黃醬是將奶黃醬煮熟，類似法國的奶黃醬，但是，這個醬更加濃稠，口感比法國奶黃醬還要更清爽，最主要一定要有肉桂與檸檬香氣，使用傳統的咖啡色粗陶土烤盤，焦糖在奶黃醬上方。

下午陽光的顏色，跟烤焦糖奶黃醬一樣，金黃色的奶霜，溫暖，又甜美。

一向喜歡牛奶，蛋黃混合液的 Lucky 忍不住的偷偷的舔了好幾口……。

加泰隆尼亞
烤焦糖奶黃醬
Crème catalane

{ **Ingredients** (Serves 2)

250 grams Milk

1 Each Cinnamon Stick

16 oz Corn flour

3 Pieces Egg Yolk

50 Gram Sugar

1 Each Lemon Rind

Dash Brown or white sugar

食材（約 2 人份）

牛奶 250 克 、肉桂棒 1 根 、
玉米粉 16 克 、蛋黃 3 顆 、
糖 50 克 、檸檬皮 1 顆 、
紅糖或是白糖少許

製作

step 1

倒出一杯牛奶，加入玉米粉稀釋攪拌，備用。

Add corn flour to 1 cup of milk. Stir to dissolve.

step 2

在鋼鍋裏放進蛋黃跟糖攪拌打混至略成淡白色。

Beat egg yolk and sugar until it turns pale white.

step 3

一支深鍋裏放入剩下的牛奶跟肉桂棒與檸檬皮，小火加熱，不要煮滾。

Add remaining milk, cinnamon stick and lemon rind to saucepan. Cook over low heat. Do not bring to a boil.

step 4

加熱牛奶慢慢倒入剛剛攪拌混合的蛋糖液裏，一邊倒一邊攪拌。

Add heated milk into egg mixture. Keep stirring when you are mixing.

step 5

在將牛奶蛋黃液在倒回深鍋裏，接着加入牛奶玉米粉混合液，拿起肉桂棒，小火煮 3~5 分鐘，直到奶黃液變稠，小心不要焦鍋。

Return egg/milk mixture into saucepan. Add corn flour mixture. Remove cinnamon stick. Cook over low heat for 3-5 minutes until mixture thickens. Be careful not to burn it.

step 6

煮到變濃稠的奶黃液到入烤盤上，放涼。

Pour thickened egg mixture into baking pan. Let cool.

step 7

要享用前，撒上紅糖，用噴火槍將糖燒成焦糖狀。

Before serving, sprinkle brown sugar on top. Torch it until it caramelized.

Helpful Tips
亮亮的小建議

1. 在步驟 5 煮奶黃醬時，很容易會焦鍋，建議一邊煮一邊使用木匙或是手動打蛋器攪拌，防止鍋底邊的奶黃醬焦掉。

2. 這道跟烤布蕾 Crème Brûle 有點不相同，在西班牙這道不需要將糖噴焦，放涼後就可以直接享用。因此，你當然也可以不用燒焦糖在奶黃醬上方。

1. *It is very easy to burn the egg mixture on step 5. Keep stirring mixture with wooden spoon or hand held mixer while cooking. This will help prevent mixture from burning.*

2. *This is a little different from Creme Brulee. In Spain, you do not have to torch the sugar until caramelized. Once it is cooled, it can be served. Therefore, you can also serve it without the caramelized sugar.*

傳統式糖烤牛奶米布丁

傳統式糖烤
牛奶米布丁
Riz au lait à
l'ancienne

西北部 諾曼第 Normandie

在路易十七時期，國王的總監長攜帶着香料跟米到法國的諾曼第區。該區的 Pays d'auge 正值鬧饑荒的時候。

"糖烤牛奶米布丁"是一道將米放在牛奶裏煮熟，能讓大家吃飽的其中一個米飯食譜。剛開始學習用牛奶來煮米飯的 Pay d'auge 居民們實在非常的飢餓，餓到無法多等待米飯的熟成，就快速的將還正在烤麵包使用的火爐裏用土碗裝滿還在烤的牛奶米飯拿出，隨即在出烤爐的米飯上發現上層一片焦黑，有些部份由於糖跟牛奶長時間煮的關係顏色轉變成帶有金黃色澤的表層，猶如路易十六時期凡爾賽宮被放火燒盡後的景象，因此，這道米飯有另了一個名稱"Tordre la goule"，中文說法是"歪扭岩石"，La goule 另外有"吸血鬼"之意，也或許可以說是當時饑餓的人民用另一種諧音取名的表現方式來取笑當時期執政的國王與皇后（是否是意指路易十六跟馬莉安東尼特皇后為吸血鬼，這就不得而知了）。在諾曼第區的方言說法則是"Teurgoule"。 無論如何，這道米飯食譜的出現，在那個時期對當地許多饑餓的人民而言正是最好的時刻。

Ingredients (Serves 3)

100 grams Round Rice

500 Ml Milk

50 grams Sugar

1 Each Vanilla Pod

食材（約 3 人份）

圓米 100 克 、牛奶 500ml 、
糖 50 克 、香草莢 1 根

Helpful Tips

亮亮的小建議

可以淋上不列塔尼特有的奶
油焦糖鹽醬或是牛奶醬，一塊
享用。

*Your can also serve it with British butter
caramel salt sauce or milk sauce.*

製作

step 1

準備一個深鍋放進米跟水，煮滾
一分鐘後瀝乾米跟水。

*Add water and rice into saucepan. Bring
to boil for about 1 minute. Drain.*

step 2

另一隻鍋裏放入牛奶＆糖，並將
香草莢對切，用刀尖將香草籽挖
出一起放進牛奶鍋裏煮。

*In another pan, add milk, sugar, vanilla
bean skin and vanilla seeds (Slice
Vanilla bean in half and scrape seeds off
the pod). Bring to a boil.*

step 3

香草牛奶糖鍋煮滾後，將香草莢
取出，放進剛剛煮過的圓米，小
火煮約 30 分鐘。

*Take vanilla bean skin out of pan. Add
the cooked rice. Cook for additional 30
minutes over low heat.*

step 4

煮的過程中，要不斷的將鍋中上
從薄膜取出。

*While it is cooking, make sure to remove
the thin white membrane off the top of
the mixture.*

step 5

煮好的牛奶米倒入烤模裏，烤箱
用 160℃ 烤 30 分鐘，米布丁烤
上色即可。

*Pour cooked rice mixture into baking
molds. Preheat oven to 160℃ . Bake
for 30 minutes until the rice pudding
browned.*

step 6

出爐後，可以熱熱的享用，也可
以冷吃！

It can be served hot or cold.

 西南部 阿基丹 Aquitaine

> *生活的小確幸，是捧着一盤阿嬤做的甜點，肩並肩坐在天氣好的搖椅上聊天談笑吃着阿嬤的家常甜點……簡單樸實的美好鄉村生活。*

糖烤牛奶燉蛋主要使用蛋跟牛奶跟糖，跟倒扣焦糖布丁十分的相似，但是，比較沒有倒扣焦糖布丁那麼滑順，主要差別在於鮮奶油的部份。糖烤牛奶燉蛋是糖在上方烤，倒扣焦糖布丁是將糖煮成焦糖放在烤模底層蛋奶液在上層。

阿基丹 Aquitaine 在法國得地理位置上比較靠近葡萄牙。葡萄牙有個較大家熟知的很著名的甜點，就是葡萄牙蛋派（亞洲稱 "葡式蛋塔"）。葡式蛋塔的內餡口感跟阿基丹的糖烤牛奶燉蛋十分接近，而阿基丹的法國糖烤牛奶燉蛋又增加了香草籽的香氣，因此，吃起來的口感上可能會在比葡式蛋塔的蛋奶液內餡在香一些，有興趣的朋友可以試試比較看看喔。

Ingredients (Serves 2)

75 grams Granulated Sugar + 1 Tbsp for decoration burnt sugar

500 Ml Milk

1 Bag Vanilla Candies

3 Each Eggs

食材（約 2 人份）

細砂糖 75 克＋ 1 湯匙裝飾烤糖用、
牛奶 500ml 、香草糖 1 包 、蛋 3 顆

製作

step 1

在深鍋裏放進糖，跟香草糖一起
煮滾。

*Add sugar and vanilla candies to sauce
pan. Bring to a boil.*

step 3

煮滾的香草牛奶糖倒進蛋液裏，
快速攪拌混勻。

*Pour vanilla candy and milk mixture
into eggs. Stir quickly to combine.*

step 5

倒進烤模裏，烤箱以 200℃ 預熱
10 分鐘烤 25 分鐘，撒上 1 湯匙
細砂糖在烤 5 分鐘即可。

*Preheat oven to 200℃ for 10 minutes.
Pour mixture into baking mold and bake
for 25 minutes. Sprinkle 1bsp of sugar
on top and bake for another 5 minutes.*

step 2

準備一個沙拉碗，將蛋放進碗裏
均勻的打散。

Beat eggs in mixing bowl.

step 4

將混合的牛奶蛋液過篩，再將泡
泡撈起。

*Pour mixture through sieve. Remove any
bubbles.*

step 6

待涼後，享用。

Cool and serve.

高亮的小建議 (Helpful Tips)

1. 如果你是使用較淺的烤模，請烤約 20~25 分鐘即可。

2. 假使沒有香草糖可以使用，可用 3/4 跟香草莢來替代。

3. 你也可以分裝在布丁罐裏進烤箱烤，烤完出來後包裝就可以讓小
   朋友帶去學校享用，方便又營養，且健康的家常甜點。

1. If you are using a shallow pan, baking time can be
   reduced to 20-25 minutes.

2. Vanilla candies can be replaced with 3/4 vanilla bean.

3. You can also bake them in small pudding pans. They
   can be wrapped when done and sent off to school with
   children. It is convenient, nutritious and healthy.

咕吉霍夫蛋糕
Kugelhopf

東北部 阿爾薩斯 Alsace

咕吉霍夫蛋糕 Kugelhopf 在法國食品歷史上很具重要意義。不論在波蘭、德國跟奧地利幾個國家都不難發現到這種事先發酵麵團做成的蛋糕。咕吉霍夫蛋糕最古老傳統的做法是用啤酒花提煉出來的啤酒酵母來製作發酵麵團，並將製作好的麵團放置在中間有凹槽像皇冠、上了釉彩後的砂土陶模裏，烤出來就會是皇冠狀蛋糕。這種蛋糕最早出現在 18 世紀鄰近德國的阿爾薩斯（Alsace）的洛林區（Lorrine）。洛林區的糕點在法國的中世紀時期早享譽盛名。路易十五國王就封給他的波蘭岳父擁有洛林及巴爾公國的君權。由於他的岳父年紀已大，所剩時間不多，因此；他將他剩下的歲月時光全揮霍在享受餐桌上的美食與豪華節慶美食上面。如咕吉霍夫這種發酵麵團的蛋糕在中世紀大多是出現在婚禮上。正由於洛林王走遍中歐各地，享遍中歐各種美食，當他發現這款味道柔軟美味的蛋糕，便將它帶回法國洛林省。某次，他的廚師為他製作這款蛋糕，並加上一杯上好的葡萄酒伴隨着蛋糕一塊品嚐，讓他為此開心了許久。之後，凡爾賽宮裏也跟着製作，像馬麗安東尼特皇后的早餐上再無法少了這款美妙味道的糕點。咕吉霍夫的商業食譜於 1840 年由喬治的糕點師傅從史特拉堡將其帶到巴黎，此一段並被記載於史學家坡耶瑜的《糕點備忘錄》裏。喬治師父並在巴黎的公雞街上開了第一家賣這個咕吉霍夫糕點的店。

{ Kugelhopf Ingredients
(250-300 grams Flour Portion)

300 grams All Purpose Flour

150 grams Milk

1 each Egg

40 grams Sugar

5 grams Sea Salt

65 grams Butter

15 grams Raisins

2 Tbsp Rum

15 Each Almond

{ Dough Starter Ingredients
(Makes a 300 grams Kugelhopf)

50 grams All Purpose Flour

10 grams Fresh Yeast

30 grams Water

咕吉霍夫甜點糕食材（約 250~300 克麵粉的蛋糕份量）

麵粉 300 克、牛奶 150 克、蛋 1 顆、糖 40 克、海鹽 5 克、奶油 65 克、葡萄乾 15 克、蘭姆酒 2 湯匙、杏仁果 15 顆

老麵種食材（約一個 300 克咕吉霍夫甜點糕用量）

麵粉 50 克、新鮮酵母菌 10 克、水 30 克

製作

製作老麵種：將麵粉、新鮮酵母菌放進一個碗裏，加入 30 克水攪拌均勻，放置室溫發酵一個小時即可使用。

Dough Starter: Put flour and yeast into bowl. Add 30 grams of water. Let rise in room temperature for one hour. It is ready for use.

step2

等待老麵發酵時間，將葡萄乾放進一個碗裏倒進蘭姆酒備用。

While dough is rising, mix raisins with rum in mixing bowl.

另外準備一個大的沙拉碗，放進已經發酵的老麵種、蛋、糖跟海鹽攪拌混合，慢慢的放進麵粉一邊攪拌一邊加入麵粉，麵粉呈現有點乾燥攪不太動時加入牛奶，將麵團跟牛奶攪拌混合，可以使用手動打蛋器輔助攪拌，再將剩下的麵粉，慢慢加入攪拌混合，這時候的麵粉會開始呈現麵團狀。

In a larger mixing bowl, combine dough starter, egg, sugar and sea salt. Slowly stir in flour, add milk when the dough becomes too dry. A hand held mixer can be used. Slowly add in all the flour. The dough will start pulling together.

step 4

奶油切小丁塊慢慢放入混合麵團,可以用手來攪拌,奶油加入後麵團會開始變得比較溼黏,偶爾手沾點乾麵粉在繼續揉混奶油與麵團。

Dice butter and mix into dough with hand. Dough will become a little sticky. Dip hand in dry flour to continue kneading the dough.

step 5

麵團與奶油混合後,加入葡萄乾,若你喜歡有酒香味可以將泡葡萄乾的蘭姆酒一併倒入麵團裏揉混合。

Once the dough and butter has been incorporated, add in the raisins. If you would like the dough to have a heavier liqueur fragrance, you can add the rum into the dough.

step 6

放置室溫一個小時。一個小時後麵團會膨脹變大。

Leave dough in room temperature for an hour. Dough will rise.

step 7

烤模塗上奶油,底層放上杏果。

Brush baking tin with butter. Layer almond in the bottom.

step 8

麵團慢慢的放進烤模裏。

Slowly put dough into baking tin.

step 9

烤箱以 180℃ 預熱 10 分鐘,烤 30~40 分鐘。

Preheat oven to 180℃ for 10 minutes. Bake for 30-40 minutes.

step 10

甜點糕出烤箱後,放涼脫膜撒上糖粉後,就可以馬上切片享用。

Once it is done, let it cook and remove from baking tin. Sprinkle icing sugar on top and serve.

Helpful Tips
亮亮的小建議

1. 這是一次可以多做幾個起來放進冷凍庫的蛋糕,要吃的時候拿出來退冰 40~50 分鐘後就能享用,不用在回溫烤過。

2. 在加入奶油的時候,建議將奶油切成小塊放進去比較好攪拌。

3. 在麵團跟奶油混合的時候,麵團會比較溼黏,建議手最後抹上乾的麵粉後在揉麵團會比較不黏手。奶油一定要融合麵團裏,烤起來才會好吃!

4. 烤的過程中,如果太快上色,可以放上一張烘焙紙減緩上色的速度。

5. 如果你不想製作老麵種可以用 20 克的乾燥麵包酵母菌替代。但是老麵種製作出來的口感比較好吃。

1. *You can make several of this dessert at the same time and freeze them. They can be defrosted 40-50 minutes before serving. It is not necessary to rebake them.*

2. *Butter should be cut up as small as possible for easier mixing.*

3. *When you mix dough with butter, the dough will become sticky. Dip hands in dry flour to ensure easier mixing of the dough. Butter has to be completely incorporated into the dough for best taste.*

4. *If the bread browns too quickly, cover the top with a parchment paper to slow down the browning.*

5. *If you do not want to make the dough starter, you can replace it with 20 grams of dry yeast. However, the dough starter has a much better texture.*

黑巧克力慕斯

Mousse au chocolat noir

全國

過簡單，美味的法式家常美味小甜點！

自己的嘴。但是，這真的是一道你絕對不能錯

濃郁，且熱量不高……大家擔心自己控制不了

不是它不好吃，相反的，是它真的太過美味、

是很多人沒有勇氣製作的一道法式家常甜點。

這是秋天餐後晚間的小小確幸。黑巧克力慕斯

吃着剛從冰箱取出奶奶下午做的巧克力慕斯，

餐後，拉一張椅子，找個靠近壁爐坐下取暖，

Ingredients (Serves 2)

2 Pieces Eggs

Dash Sea Salt

2 Tbsp Granulated Sugar

100 Gram Dark Chocolate

25 grams Butter

食材（約 **2** 人份）

蛋 2 顆、海鹽小撮、細砂糖 2 湯匙、黑巧克力 100 克、奶油 25 克

製作 —————————————————————————

step 1

將蛋白＆蛋黃分開。

Separate egg white and egg yolk .

step 2

蛋白加海鹽用電動打蛋器打到蛋白可以在攪拌器上不會下垂或是掉下來的程度，放進冰箱冷藏，備用。

Add sea salt to egg white. Beat with electric mixer until stiff peaks formed. Refrigerate.

step 3

蛋黃＆糖一起打到呈淡黃色。

Beat egg yolk with sugar until pale yellow.

step 4

一個鍋子加入水，放在火爐上，另外一個鋼鍋將巧克力敲碎放入加入少許水，隔水加熱方式將巧克力融化。巧克力融化之後，離開火源，加入切塊的奶油跟融化的巧克力攪拌混合均勻。

Add water to saucepan and put it on the stove. Add chopped chocolate with a little water in a stainless steel bowl. Put bowl over saucepan and melt chocolate in the water bath. Once chocolate is melted, remove from heat. Add cut up butter. Mix until combined.

step 5

接着將蛋糖液加入融化巧克力裏面，再度慢慢的將他們攪拌混合。

Add egg sugar mixture into chocolate. Mix until combined.

step 6

從冰箱裏拿出剛剛打發的蛋白，加入融化巧克力跟蛋糖混合液裏，再將他們攪拌混合到完全看不到蛋白，完全的融合一起。

Take beaten egg white out from fridge. Gently fold into chocolate/egg mixture. Mix until all the egg white has been incorporated into the chocolate mixture.

step 7

倒入杯子裏，送進冰箱至少 3 個小時後享用。

Divide into cups. Refrigerate at least 3 hours before serving .

法國土司
烤蘋果

Pain perdu au
pommes rôties

全國

以前法國家庭會經常做"煎奶油焦糖蘋果"放在法國麵包上享用。而這個麵包就會自然的會吸收這奶油焦糖蘋果的湯汁，香濃奶油香氣與蘋果果汁融合一起那麼美好的醬汁讓許多貪好美食的老饕很愛這個味道。這個食譜轉變來自煎奶油焦糖蘋果，比較不那麼多厚重奶油，也保有奶油焦糖香，蘋果直接烤過，更可以品嚐到蘋果原味。假日的早晨就是要慵懶的起床，吃美味百分百早午餐"法國土司烤蘋果"跟一杯熱熱的公平交易咖啡……假日就是要慢慢的享受這一切的美好……。

Ingredients (Serves 4)

4 Each Small apples

4 Each Country Style bread or Butter Bread

2 Each Eggs

4 Scoops Vanilla Ice Cream

50 grams Pistachio

1 Cup Milk

Syrup

125 grams Sugar

125 grams Butter

1 Cup Apple Liqueur

食材 (約 4 人份)

小蘋果 4 顆、鄉村麵包或是奶油麵包 4 片、蛋 2 顆、香草冰淇淋 4 球、開心果仁 50 克、牛奶 1 杯

糖漿

糖 125 克、奶油 125 克、蘋果酒 1 杯

製作

step 1

深鍋裏放入糖跟 125 克的奶油慢火煮融化，趁熱加入蘋果酒，小火慢慢煮到變糖漿的濃稠度，放涼，備用。

Heat sugar and 125 grams of butter until butter dissolves. Add apple liqueur when mixture is still hot. Slowly cook until syrup thickens. Cool and set aside .

step 2

蘋果洗淨後，將蘋果頭以平切切開。去掉中心的梗跟籽，放進烤箱 180℃烤 20 分鐘。

Wash apples and slice apple tops off horizontally. Remove seeds and core. Bake at 180℃ oven for 20 minutes

step 3

烤蘋果的時間，準備一個平底鍋加熱放一點點奶油進去，將牛奶跟蛋放在一個沙拉碗裏打散，並且混合。放進麵包沾取蛋牛奶液，接着放入平底鍋裏間上色（中小火）。

When apples are baking, heat a little butter in a frying pan. Add milk with eggs to mixing bowl. Dip bread in egg mixture. Pan fry on pan at medium heat until browned.

step 4

在換面煎上色取出放在盤上。

Flip bread over and brown on the other side. Set onto serving plate.

step 5

蘋果出爐放在熱麵包上面，挖取香草冰淇淋填進蘋果芯裏，撒上些許開心果碎，蓋上蘋果頭，淋上蘋果酒奶油焦糖醬汁再撒上一些開心果碎裝飾。

Lay baked apples on top of hot bread. Scoop ice cream into apple. Sprinkle chopped pistachios on top. Cover with apple lids. Pour apple liqueur butter caramel sauce on top and sprinkle additional pistachio on top.

Helpful Tips
亮亮的小建議

1. 法國吐司用的麵包最好選擇已經久放的麵包。久放的麵包浸泡在蛋奶液裏的時間需要比一般的麵包久一點，這是方便讓蛋奶液可以浸入到麵包內層裏，使原本硬硬的麵包變得更加柔軟。

2. 挖蘋果時，可以用喝茶的湯匙輔助挖出蘋果芯，但千萬別將蘋果底層跟外表挖破喔。

3. 蘋果酒奶油焦糖醬可以在前一天先製作好，或是一次製作多量，放冰箱，建議使用：淋在冰淇淋或是優格上一塊享用。

1. *It is better to use aged bread. Aged bread will absorb more egg mixture and makes the bread really soft.*

2. *You can use a teaspoon to help scoop out the seeds and cores. Make sure you do not break the bottom of the apples.*

3. *Apple butter sauce can be made a day ahead. You can also make a larger quantity and refrigerate the unused portion. It is also good with ice cream and yoghurt.*

甜橙
巧克力慕斯
- - - - - - - - - -
Mousse au
chocolat

法國老一輩的人,非常喜愛歡在做巧克力慕斯同時加些香氣進去,甜橙就是其中一種水果香味,也是老少咸宜的大家都能夠接受的 水果口味。製作巧克力慕斯時,若能夠加入甜橙果肉那就更加的完美了! 選擇巧克力的時候,建議可以選用有核仁果豆的巧克力,或是選用也可以用黑巧克力。可以的話,蘭姆酒換成白蘭地酒 也會有很不錯的酒香口感,加入酒香的慕斯會是個適合冬天的巧克力甜點。若是不喜歡有酒香味,不妨可將蘭姆酒替換成甜橙糖漿或是其他水果糖漿就會適合小朋友來品嚐。

有涼意的夜晚,小飲一小杯葡萄牙帶回的白蘭地之後,品嚐甜橙巧克力慕斯……有暖意的夜秋～。

Ingredients (Serves 4)

1 Each Sweet Orange

125 grams Dark Chocolate

30 grams Room Temperature Butter

2 Tbsp Honey

1 Tbsp Rum

125 grams Whipped Cream

Dash Sweet Orange Rind

食材 (約 4 人份)

甜橙 1 顆、黑巧克力 125 克、室溫融化奶油 30 克、蜂蜜 2 湯匙、蘭姆酒 1 糖匙、霜狀鮮奶油 125 克、甜橙皮 少許

 製作

step 1

甜橙去皮,去籽,果肉切塊。

Remove skin and seeds from orange. Cut into chunks.

step 2

將巧克力、蘭姆酒,蜂蜜跟奶油放進鍋子裏,放上火爐上小火將巧克力跟奶油煮化。

Add chocolate, rum, honey and butter to saucepan. Cook over low heat until chocolate and butter melt.

step 3

放入四湯匙的霜狀鮮奶油跟甜橙塊,攪拌混合後離火。

Add 4 tbsp of whipped cream to oranges. Stir to combine. Remove from heat.

step 4

裝進杯子裏,放進冰箱至少 3 個小時。出冰箱享用前,放上一小匙的霜狀鮮奶油跟甜橙皮做裝飾即可。

Divide mixture into cups. Refrigerate at least 3 hours before serving. Add a dollop of whipped cream and orange rind on top to serve.

part 4

法國家常
鹹 / 甜蛋糕

甜橙蛋糕
Gâteaux à l'orange

這是阿嬤的拿手蛋糕之一，以前我在亞洲吃到的蛋糕大多偏乾，品嚐的過程需要藉由不斷的喝水改善蛋糕讓嘴巴乾渴的情況。

阿嬤的這個甜橙蛋糕的蛋糕體比較為溼潤，加上使用新鮮甜橙汁製作，吃起來會有如喝到新鮮甜橙果汁的感覺。溼潤的蛋糕即使一次吃不完放在冰箱都不會因為放置時間過久而導致蛋糕失去水分造成蛋糕體太乾的現象。

一時間買太多柑橘水果⋯⋯吃不完呢？來做做簡單，美味又快速的甜橙蛋糕吧～只要 30 分鐘的時間就能完成囉！

{ Ingredients

2 Each Sweet Oranges

1 Each Lemon (optional)

2 Pieces Eggs

115 grams Melted Butter

175 grams Icing Sugar

115 grams All Purpose Flour

11.5 grams Baking Powder

食材

甜橙 2 顆、檸檬（不是必須的）1 顆、蛋 2 顆、融化奶油 115 克、糖粉 175 克、麵粉 115 克、泡打粉 11.5 克

製作

step 1

拋出檸檬碎皮＆甜橙碎皮。

Scrape orange and lemon rinds.

step 3

之後加入麵粉過篩和泡打粉。將
甜橙皮切細或拋碎加入，可以的
話再加入半顆檸檬皮。擠出甜橙
汁加入麵糊裏。

*Add sifted flour and baking powder. Add
orange rind and half of the lemon rind.
Juice the sweet orange and add to batter.*

step 5

甜橙糖漿製作：以半顆甜橙汁＋
剩下 50 克的糖煮成糖漿

*Sweet Orange Syrup: Use juice of half
the sweet orange + remaining 50 grams
of salt to make syrup.*

step 2

混合融化的奶油，115 克的糖粉
和蛋攪拌均勻。

*Mix butter, 115 grams of icing sugar
and eggs.*

step 4

烤箱以 200℃ 預熱 10 分鐘。將
奶油塗在烤模上，在撒上一些麵
粉。方便等一下脫膜。將剛剛預
拌好的麵糊倒入烤模裏烤個 20
分鐘。

*Preheat oven to 200℃ for 10 minutes.
Brush butter onto baking tin. Also
sprinkle flour for easy release. Add
batter into tin and bake for 20 minutes.*

step 6

淋上甜橙糖漿，即完成了。

Pour orange syrup on top. Serve.

Helpful Tips 亮亮的小建議

1. 甜橙糖漿煮好後放涼，蛋糕熱熱烤好出爐要馬上淋上糖漿，這樣
蛋糕才會有市售蛋糕一樣 表層有光澤感。

2. 吃不完的蛋糕，放冰箱可以放一週，請盡量裝進保鮮盒裏在放進
冰箱冷藏。

3. 送人時 建議可以倒入紙模烤杯裏烤，表層塗上甜橙糖漿外，你也
可以擠上法式香堤（做法在法國基本功夫六 -6）放進蛋糕盒裏，
送人時很漂亮喔！

1. *Cool orange syrup when done. Once cake is done, pour
syrup over immediately. This way, the cake will have
the same glaze as store bought cakes.*

2. *Leftover cakes can be stored in fridge for one week.
Put leftovers in covered box for refrigeration.*

3. *If it is a gift, I suggest baking them in paper cups.
Brush top with orange syrup and decorate with
Chantilly cream. (French basic technique 6). It is very
nice and presentable in a cake box.*

紅蘿蔔葡萄乾果仁蛋糕

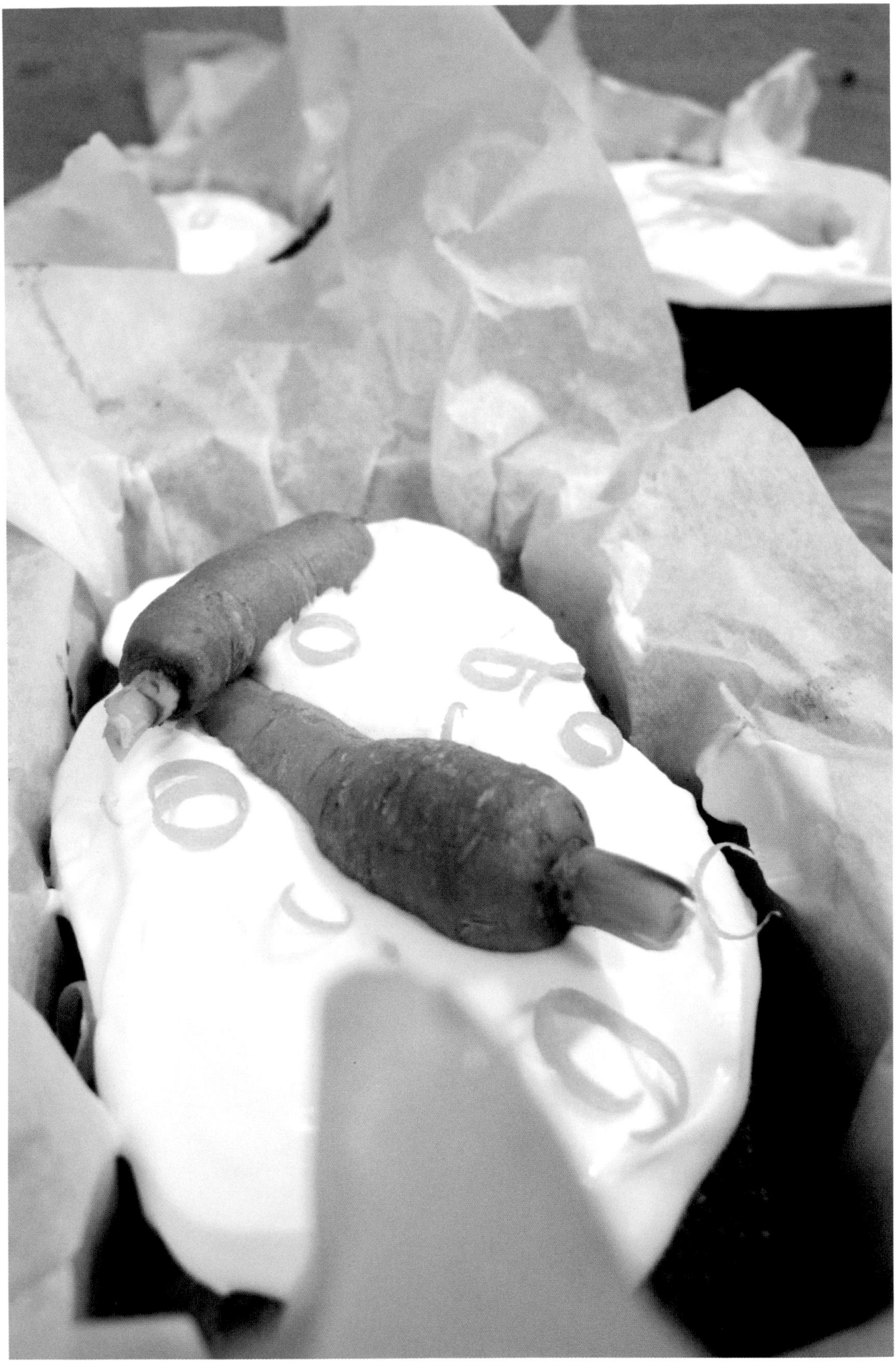

紅蘿蔔葡萄乾
果仁蛋糕

Gâteaux
aux carottes

每年菜園裏的紅蘿蔔收成時，總是會有吃不完的紅蘿蔔，那年剛好遇到種在後院裏好幾年都沒有什麼結果的核桃，剛好有結果收成，奶奶為了解決家裏有不愛吃紅蘿蔔的孫子跟吃不完的紅蘿蔔的小煩惱，自己試做了好多次不同版本的紅蘿蔔核桃蛋糕，才得到孫子們心滿意足的做法，還將蛋糕全吃光，直嚷嚷好好吃！此後，每年夏天紅蘿蔔收成時，是這些孫子們期待再度品嚐自家種的紅蘿蔔做成的蛋糕。

> 想念起奶奶溫暖的手，在廚房裏去着紅蘿蔔皮，轉身笑笑對身後的一群吵鬧的小毛頭們說
> "快了 在等等喔，好吃的蛋糕快好了" 簡單的一句話，卻看到幸福洋溢在奶奶的臉上。

{ Ingredients (Serves 10)

175 ml Sunflower Oil

175 grams Brown Sugar

3 Each Eggs

85 grams Raisins

175 grams Shredded Carrots

55 grams Almond or Walnut

175 grams All Purpose Flour

1 tsp Baking Soda

1/2 tsp Nutmeg Powder

Dash Sweet Orange Shreds

2~3 Each Tiny Carrots (optional)

{ Decorating French Orange Chantilly Frosting

150 grams Whipping cream

70 grams Icing Sugar

2 Tbsp Sweet Orange Juice

食材（約 10 人份）

食用葵花油 175ml、紅砂糖 175 克、蛋 3 顆、葡萄乾 85 克、紅蘿蔔絲 175 克、杏仁果或是核桃 55 克、麵粉 175 克、食用小蘇打粉 1 茶匙、
肉豆蔻粉 1/2 茶匙、甜橙絲少許、小根紅蘿蔔（不是必須的）2~3 根

裝飾法式甜橙香堤

液態鮮奶油 150 克 、糖霜 70 克、甜橙汁 2 湯匙

step 1

準備一個沙拉碗，放進食用葵花油、糖跟蛋攪拌混合，接着放入紅蘿蔔絲、葡萄乾，跟果仁與甜橙皮。

Combine sunflower oil, sugar and eggs in mixing bowl. Add carrots, raisins, nuts and orange rinds.

step 2

將麵粉、小蘇打粉、肉豆蔻粉過篩之後加入混合攪拌均勻。

Mix sifted flour, baking soda, and nutmeg powder.

step 3

將烤模塗上奶油，放上一張烘焙紙，倒入麵糊。烤箱以 180℃ 預熱 10 分鐘烤 40~45 分鐘。

Brush mold with butter. Cover with parchment paper. Pour batter in. Preheat oven to 180 for 10 minutes. Bake for 40-45 minutes.

step 4

準備一鍋水，小根紅蘿蔔水煮煮熟後瀝乾水份，備用。

Prepare a pot of hot water. Boil mini carrots until cooked. Drain and set aside.

step 5

烤箱在烤蛋糕的同時，準備製作法式甜橙香堤醬，準備一個沙拉碗，放入鮮奶油、糖與甜橙汁，用電動打蛋器打到鮮奶油變成霜狀即可（相同做法可參考 法國基本練功夫 六 -5）

While cake is baking, prepare French orange frosting. Add whipping cream, sugar and orange juice to mixing bowl. Beat with electric mixer until mixture thickens and is creamy. (please refer to French Basic Technique 6-5)

step 6

烤箱拿出烤好的紅蘿蔔蛋糕，放涼之後，塗上剛剛打好的甜橙香橙醬放上小根紅蘿蔔或是甜橙絲亦可。

Cool carrot cake. Spread frosting on top. Decorate with mini carrots and orange shreds.

Helpful Tips
亮亮的小建議

1. 如果不用烤膜來烤蛋糕，可以考慮用紙烤模裏面放進一張烘焙紙在倒入麵糊，進烤箱烤，之後塗上香堤醬撒上甜橙絲，很漂亮，可以一次放兩三個在蛋糕盒裏，送人也很大方喔！

2. 食用小蘇打粉可以以無鋁泡打粉來取代

3. 喜歡黑糖的朋友，可以將 175 克的紅砂糖以同比例的黑糖取代喔！口感更香，甜度也比較不甜。

4. 肉豆蔻可以到中藥房購買到整顆的比較香也比較便宜，買回家在用磨一磨出粉即可，或是到大的有賣進口商品的超市買整罐的肉蔻粉亦可，但是比較貴。

1. *Instead of a large baking mold, you can also bake in smaller paper molds lined with parchment paper. Spread orange Chantilly dressing on the cakes and sprinkle with sweet shredded orange. You can put 2 to 3 smaller size cakes into the cake box. They are very presentable.*

2. *Baking soda can be replaced with aluminum free baking powder.*

3. *Brown sugar can also be replaced with dark brown sugar to increase fragrance and reduce sweetness.*

4. *Grinding whole nutmeg purchased from herbal store is preferable because of the cost and fragrance. However, if those are not available, ground nutmeg can be purchased at supermarkets at a higher cost.*

薇多麗雅
戚風蛋糕
Génoise Victoria

歐洲大多數的家庭都會做這個蛋糕，其主要以戚風蛋糕為蛋糕基體，果醬，法式香堤跟草莓組合而成的家常蛋糕。

這個蛋糕有個夢幻且富麗堂皇印象的名稱，蛋糕裏面的草莓與法式香堤都是很受歐洲女性鐘愛的甜點元素，因此，另外有公主蛋糕的稱號。又因為這個蛋糕在英國發跡，所以在法國便有英國蛋糕的稱呼。

我的住家離英國算相當近，搭船即可到達，也常常有英國人來法國小旅行，這個薇多麗雅蛋糕便成為我們這一區的家常小蛋糕之一。
炙熱陽光的下午，跟小姪女們採草莓去吧！

{ Ingredients (Serves 8)

175 grams Room Temperature Softened Butter

175 grams All Purpose Flour

1 Tsp Aluminum Free Baking Powder

175 grams Brown Granulated Sugar

3 Each Eggs

Dash Icing Sugar

3 Tbsp Strawberry or Raspberry Jam

{ Filling Ingredients

300 ml Fresh Cream

30 grams Icing Sugar

16 Each Medium Strawberries

食材（約 8 人份）

室溫融化奶油 175g、麵粉 175g、無鋁泡打粉 1 茶匙、紅細砂糖 175g、蛋 3 顆、糖霜少許、草莓果醬或是覆盆子果醬 3 湯匙

內餡材料

鮮奶油 300ml、糖霜 30 克、中型草莓 16 顆

製作

step 1

烤箱以 180℃ 預熱 10 分鐘，烤模底層與四周塗上奶油跟撒上許的麵粉、方便蛋糕脫膜。

Preheat oven to 180℃ for 10 minutes. Brush baking tin with butterl and sprinkle with flour for easy release.

step 2

麵粉與泡打粉一起過篩在一個鋼鍋裏，加入奶油。糖與雞蛋。

Sift flour and baking powder into mixing bowl. Add butter, sugar and eggs.

step 3

用打蛋器打混所有的材料變成麵糊。

Use electric mixer to blend the batter.

step 4

將麵糊倒入烤模裏，烤 25~30 分鐘，使用牙籤插入蛋糕體裏若是牙籤乾燥的表示蛋糕可以出爐了。

Pour batter into baking tin and bake for 25-30 minutes. Stick toothpick into cake. If toothpick comes out clean, cake is done.

step 5

蛋糕出爐後，放冷約 5 分鐘的時間，在蛋糕的橫面對切成兩塊。

Cool cake for 5 minutes after it is done. Slice cake horizontally into half.

step 6

準備另一個鋼鍋，倒入鮮奶油，使用電動打蛋器用低速打發鮮奶油，加入糖霜後在用高速將鮮奶油糖霜打發成霜狀，即變成是法式香堤即可，將草莓去掉綠蒂後對切。

In another mixing bowl, beat whipping cream at low speed with electric mixer. Add icing sugar and continue to beat at high speed until it thickens. This is French Chantilly Cream. Remove stems on strawberries. Cut in half vertically.

step 7

蛋糕對切底層那塊塗上果醬，在放上法式香堤在，擺上法式香堤上方，擺上對切的草莓。蓋上另外一塊切開的蛋糕，撒上糖霜，就可以切片享用囉！

Spread bottom layer of cake with jam. Follow with a layer of the Chantilly cream. Layer cut strawberries on top. Cover with top layer of cake. Sprinkle with icing sugar. Slice and serve.

Helpful Tips
亮亮的小建議

你也可以像我的小姪女一樣在底層蛋糕塗上果醬後，放上草莓，在塗上法式香堤後再一層草莓，切片後也可以在蛋糕頂端放上一小跎法式香堤在擺上一顆草莓。既然都是說是公主蛋糕了，當然沒有什麼不能的囉。

You can also be creative like my niece. After the jam on the bottom, she layered strawberries on top followed by Chantilly cream and then another layer of strawberries.

You can also slice it and then add a dollop of Chantilly cream and additional strawberries on top. Since this is a Princess cake, the possibilities are unlimited.

椰絲香蕉蛋糕

椰絲 香蕉蛋糕
Cake à la banane et à la noix de coco

阿嬤就像一座寶藏一樣，總有源源不斷的甜點食譜，當有時候，手邊的材料有限，阿嬤會像魔術師一樣地馬上告訴我如何用現有的材料來做什麼樣的甜點，當然，這些糕點的做法都不是她憑空想像來的，阿嬤說，以前她的媽媽總會做出一些好吃的甜點給她們家中的小孩吃，但是，以前的鄉下地方不像大城市那樣可以任意的採買要做的甜點材料，她的媽媽就會利用家裏剩下的材料來做出一樣好吃的糕點讓他們享用。椰子絲香蕉蛋糕，就是阿嬤的媽媽傳授的糕點裏其中一個私房食譜。阿嬤說她的媽媽是以製作做鹹蛋糕的原理來製作這道甜品，原本這道蛋糕應該用霜狀鮮奶油來製作，剛好家裏沒有霜狀鮮奶油，就用酸奶油替代，沒有想到一樣好吃！香蕉的部份可以任意換成你喜歡的水果泥，就會成為一道法式家常很隨意自由變化的蛋糕。

{ **Ingredients** (Makes 2 cakes)

90 ml Olive Oil

11/2 tsp Aluminum Free Baking Powder

250 grams All Purpose Flour

200 grams Sugar

55 grams Shredded Coconut

2 Each Eggs

2 Each Ripe Bananas (mashed)

125 ml Sour Cream

1 Tsp Vanilla Essence

食材 (約兩份蛋糕)

橄欖油 90ml、無鋁泡打粉 1 1/2 茶匙、麵粉 250 克、糖 200 克、椰子絲 55 克、蛋 2 顆、熟成香蕉 (事先搗碎) 2 條、酸奶油 125ml、香草精 1 茶匙

製作

step 1

先將麵粉跟泡打粉過篩。

Sift flour and baking powder .

step 2

加入椰子絲跟糖，稍加攪拌混合。

Add coconut and sugar. Mix lightly.

step 3

另一個鍋子裏打入兩顆蛋，一邊加入橄欖油一邊攪拌混合油跟蛋，之後加入香草精跟酸奶油，再度混合攪拌均勻。

In another bowl, add eggs and olive oil. Stir until combined. Continue to add vanilla essence and sour cream. Mix until combined.

step 4

將蛋，橄欖油的混合液體加入剛剛已經混合攪拌的麵粉，椰子絲跟糖裏，在一次攪拌混合後加入香蕉泥，只要拌勻即可。

Add egg mixture to the flour mixture. After all ingredients had been combined, add mashed banana. Mix until just combined.

step 5

烤模塗上橄欖油或是奶油，從烤模底部跟四周都塗上奶油，之後倒入麵糊。

Brush baking tin with olive oil or butter. Pour in batter.

step 6

烤箱以 180℃ 預熱 10 分鐘烤 1 小時。

Preheat oven to 180℃ for 10 minutes. Bake for an hour.

step 7

出烤爐後，等蛋糕涼之後脫烤膜，就可以切片享用了。

When cake is done, let it cool and then remove from baking tin. Slice and serve.

Helpful Tips
亮亮的小建議

1. 橄欖油可以用葵花油來替代。

2. 酸奶油的部份可以選用法國霜狀鮮奶油來製作一樣美味喔！

3. 香草精只要到一般的烘焙店都可以買得到唷！有興趣的也是可以自己製作，用起來會更佳安心。

4. 如果可以找到新鮮的椰子絲風味會更好，若是找不到椰子粉或末也都可以的。

5. 如果蛋糕切開看到蛋糕體內感覺還是有點濕潤 不用太擔心 有可能是你的香蕉搗的不夠碎有塊狀。這是可以接受正常的狀態，香蕉的塊狀越大 看到蛋糕溼度的部份越多。冷卻後就好了！

1. Olive oil can be replaced with Sunflower oil.

2. Sour cream can be replaced with French whipped cream.

3. Vanilla essence can be purchased at most baking stores. If you like, you can also make it yourself.

4. Fresh shredded coconut are more desirable if you can find them. Otherwise, dried coconut shreds or powder are fine.

5. If the cake looks wet once it is cut, dont be too worried. It is probably because the banana was not mashed enough and there are some bigger pieces. This is an acceptable normal condition. The larger the banana pieces are, the more this will happen. Once the cake cools down, it will be OK.

濃 郁的巧克力醬蛋糕，濃郁、甜而不膩的巧克力加上鬆軟的蛋糕，讓人實在很難拒絕它！ 相信我吃完後，你會像我一樣 不自覺的開心起來，手舞足蹈，跟着音樂搖擺着你的身體……。

濃情 巧克力醬蛋糕

Gâtaeu au chocolat
neppé de fudge

巧克力蛋糕是在法國每個家庭裏無敵超級受歡迎的糕點其一種，家裏不論任何誰的生日或是法國節慶，多少都會出現一個巧克力蛋糕，不管論是大人，還是小朋友都會吃得滿嘴的巧克力醬在嘴邊四周。

Ingredients (Serves 4)

88 grams Butter

88 grams Fine Brown Sugar

1.5 Tbsp Maple syrup or honey

2 Each Eggs (beaten)

20 grams Almond Powder

88 grams All Purpose Flour

3 grams Aluminum Free Baking Powder

Dash Sea salt

20 grams Cocoa Powder

Chocolate Sauce

180 grams Dark Chocolate

28 grams Brown Sugar

180 grams Butter

2.5 Tbsp Extended Shelf Life Milk

1/2 Tsp Vanilla Essence

食材（約 4 人份）

奶油 88 克、細紅砂糖 88 克、楓糖漿或是蜂蜜 1.5 湯匙、蛋（事先打散）2 顆、

杏仁粉 20 克、

麵粉 88 克、

無鋁泡打粉 3 克、

海鹽一小撮、

可可粉 20 克

巧克力醬

黑巧克力 180 克、

紅砂糖 28 克、

奶油 180 克、

保久乳 2.5 湯匙、

香草精 1/2 茶匙

製作

step 1

烤模塗上奶油後撒上可可粉或是剪一張與烤模底部一樣大小的烘焙紙，包括烤模邊邊也貼上烘焙紙。

Brush baking tin with butter and sprinkle with cocoa powder or cut a piece of parchment paper to cover the side and bottom of tin.

step 2

鋼鍋裏放入糖跟蛋，電動打蛋器將蛋糖打至略微有點淺白色後加入蛋液，糖漿跟杏仁粉，再一次將蛋糊打混合。

Cream butter and sugar with electric beater until it turns pale white. Add beaten eggs, syrup and almond flour. Mix well.

step 3

麵粉、泡打粉、海鹽過篩加入攪拌後，再將可可粉過篩，再度攪拌均勻一次。

Sift flour, salt and baking powder. Mix into butter mixture. Add cocoa powder. Mix again.

step 4

烤箱以 180℃ 預熱 10 分鐘，混合好的麵糊倒進烤模後，進烤箱烤 30 分鐘，若上色過快放上一張烘焙紙，防止過度上色。

Preheat oven to 180℃ for 10 minutes. Pour batter into baking tin and bake for 30 minutes. If batter browns too quickly, cover it with parchment paper to slow it down.

step 5

融好巧克力醬後，放進冰箱 1 小時備用。

Melt chocolate sauce. Refrigerate for one hour before use.

step 6

烤好的蛋糕，放冷後對切。底層蛋糕塗上巧克力醬，在蓋上另一片巧克力蛋糕。在塗上剩下的巧克醬，蛋糕上層稍微抹平，再將巧克力蛋糕周圍一圈塗上剩下的巧克力醬，就可以馬上切片享用。

Cool baked cake. Slice cake in half horizontally. Spread chocolate sauce on bottom layer. Cover with top layer and spread with chocolate sauce. Ice side of cake with remaining sauce. . Smooth icing around side and top of cake. Slice and serve.

1. Fudge 我們暫時稱呼它為 "富奇"，這是一種巧克力醬，主要以黑巧克力跟保久乳與牛奶組合混合而成的巧克力醬。與甘那許巧克力淋醬只有差別在加入牛奶的部份。Fudge 富奇巧克力醬在法國甜點裏會被運用在蛋糕夾層或是表面裝飾。我個人喜歡不是很甜的甜點，因此，我選擇用紅砂糖來替代白細砂糖去做蛋糕體跟巧克力醬，楓糖漿也可以用蜂蜜或是金色糖漿來替代除了增加香氣也增加些許的甜度。

2. 亞洲天氣過熱，Fudge 巧克力醬塗上蛋糕會很快融化，因此，做好的蛋糕要盡快進冰箱冷卻。冰太久的巧克力蛋糕也會過硬，所以，想要食用時先切一塊出來，等待巧克力稍微回溫再享用，會更美味。巧克力蛋糕必須在冷藏室存放可以放上 10~12 天的時間。

1. *Fudge is a type of chocolate sauce. Its main ingredients are dark chocolate, extended shelf life milk and milk mixture. The difference between fudge and ganache is the milk portion. Fudge is often being used for filling and icing of cakes in French desserts. Since I have a preference for a less sweet cake and chocolate sauce, I have opted to use brown sugar instead of white sugar. Maple syrup can also be replaced with honey or golden syrup . It is sweeter and more aromatic.*

2. *Since it is overly warm in Asia, fudge spread on cakes will melt quickly. Therefore, once the cake is done, it should be refrigerated as soon as possible. However, if the cake is chilled for too long, it will be hard. Therefore, when you are ready to consume the cake, you should cut a slice and let the chocolate sauce thaw for a while to give it the optimum taste Chocolate cake can be refrigerated up to 10-12 days. .*

鬆軟的杏仁蜂蜜蛋糕，聞起來有那淡淡的蜂蜜香，好適合來杯春茶，一塊享用。

這道糕點是法國一般家庭時常做的家常小蛋糕，很容易製作，祕訣就是將糖漿的部份煮熱後，加入麵粉跟蛋和泡打粉，就能烤出好吃的小蛋糕。

杏仁蜂蜜蛋糕
Gâteau au miel et aux amandes

我們很多蛋糕都用紅細砂糖來製作，比較不那麼甜，又可以增加一些蔗糖香氣。如果，你想要再減低糖份，在烤完後，淋上熱蜂蜜這個步驟就省略掉吧！即使省略淋上熱蜂蜜也不減蛋糕的美味喔。建議：如果不使用熱蜂蜜淋蛋糕，那就考慮擠上一球自己現打得"法式香堤醬"馬上就會變成另一種法式的杯子小蛋糕囉！

Ingredients (Serves 4)

75 grams Butter

60 grams Fine Brown Sugar

88 grams Honey

1 Tbsp Lemon Juice

1 Each Egg (beaten)

100 grams All Purpose Flour

5 grams Baking Powder

5 grams Sliced Almond

食材（約 4 人份）

奶油 75 克、紅細砂糖 60 克、蜂蜜 88 克、檸檬汁 1 湯匙、蛋，事先打散 1 顆、麵粉 100 克、泡打粉 5 克、杏仁片 5 克

製作

烤箱以 180℃ 預熱 10 分鐘。

Preheat oven to 180℃ for 10 minutes.

step 2

將奶油，糖跟蜂蜜和檸檬汁放進一個深鍋裏，放上火爐，小火將糖漿煮有點稠度，千萬不要煮滾了！

Add butter, sugar, honey and lemon juice into saucepan. Heat over low heat until slightly thickened. Be careful not to burn it!

step 3

糖漿離開火爐後，快速的加入打散的蛋，攪拌混合，接着麵粉跟泡打粉過篩後加入後再度攪拌均勻，這時候會變成有點濕濕的麵糊狀。

Remove syrup from heat. Quickly add in beaten egg. Mix. Add sifted flour and baking powder. Stir to combine. Batter will be wet.

step 4

倒入烤模裏約 7 分滿，在撒上杏仁片在麵糊上。

Pour batter into baking mold, about 70% full. Sprinkle sliced almond on top.

step 5

送進烤箱烤 30 分鐘。烤好出烤箱，另外，你可以將蜂蜜用隔水加熱的方式將蜂蜜加熱，等蛋糕烤出爐後淋上蛋糕之後，享用。

Bake for 30 minutes. Your can also melt honey over hot water and pour onto cake once it is done. Enjoy!

聖誕節過後，在後院核桃樹下撿拾核桃，今年春天，正好可以做核桃咖啡蛋糕捲，這是一道屬於春天的蛋糕……。鎮上許多戶人家的庭院都種上一兩棵核桃樹，每到聖誕節過後，核桃成熟後落果在潮濕的草地上，拾起成熟落下的核桃，好好保存可以用一兩年。

咖啡核桃
蛋糕卷

Roulé au café
et aux noix

食材（約 6 人份）
蛋 3 顆、細紅糖 115 克、濃縮咖啡 1 湯匙、麵粉過篩 75 克、核桃敲碎 30 克、濃縮鮮奶油 175 克、糖粉 40 克、濃縮咖啡 2 湯匙

{ **Ingredients** (Serves 6)

3 Each Eggs

115 grams Fine Brown Sugar

1 Tbsp Concentrated Coffee

75 grams Sifted All Purpose Flour

30 grams Chopped Walnut

175 grams Concentrated Whipping Cream

40 grams Icing Sugar

2 Tbsp Concentrated Coffee

step 1

用深鍋煮一鍋熱水，離開火源，趁水仍滾燙時，鋼鍋裏放進蛋跟糖，使用電動打電器打發到蛋糖成慕斯狀，即可。

Boil a pot of water. Remove from heat. While water is still hot, add eggs and water to stainless steel mixing bowl. Place bowl on top of hot water and whisk with electric mixer until it resembles mousse.

step 2

加入咖啡，過篩麵粉跟核桃，攪拌混勻。

Add coffee, sifted flour and walnut. Mix until combined.

step 3

烤盤上放上一張烘焙紙，塗上少許的橄欖油或是奶油。

Cover baking pan with parchment paper. Brush lightly with olive oil or butter.

step 4

攪拌好的麵糊倒進烤盤，烤箱180℃預熱10分鐘烤12~15分鐘。

Pour batter into baking pan. Preheat oven to 180℃ for 10 minutes. Bake for 12-15 minutes.

step 5

另一鋼鍋放進濃縮鮮奶油，咖啡跟糖粉 打至變成濃稠霜狀放進冰箱，備用。

Add concentrated whipping cream, coffee and icing sugar into another mixing bowl. Beat until thickened. Refrigerate for later use.

step 6

蛋糕烤好後先切掉四邊，換一張烘焙紙，趁蛋糕還熱將蛋糕捲起來。

After cake is done, trim off all four sides. Use another parchment paper and roll up cake while it is still hot.

step 7

蛋糕涼後，在打開塗上剛剛打好的奶油霜，再將蛋糕捲起來即可。撒上糖粉即可享用。

When cake cooled, open it up and spread with whipped cream. Roll cake up again and sprinkle with icing sugar. Serve.

Helpful Tips

亮亮的小建議

1. 使用濃縮咖啡蛋糕咖啡的味道不會那麼濃郁，若要咖啡香氣濃郁可以考慮咖啡酒或是咖啡精。

2. 核桃放進麵糊裏時一定要敲碎，不能像我一樣食心，放太多大顆的捲的時候，有顆粒的位置不好捲會裂開。核桃敲碎入捲的時候會比較漂亮。

3. 一定要等蛋糕體冷卻再塗上奶油霜，蛋糕體太熱塗上奶油霜會化掉變成液體。

4. 也可以在蛋糕裏的3顆蛋＋糖混合的部份另外再多加一顆蛋白一起打發，蛋糕體會更綿密。

1. When using concentrated coffee, the coffee taste is not very strong. If you want to have a stronger taste, you can consider using coffee liqueur or coffee essence.

2. Walnuts had to be chopped into small pieces before mixing into the batter. If you are too greedy and added too many large pieces, the cake may be difficult to roll and the large pieces will also cause the cake to crack. Smaller pieces will improve the appearance of the cake.

3. Make sure cake is completely cooled before spreading the whipped cream on top. Otherwise the cream will melt.

4. Adding an extra egg white to the 3 eggs + sugar mixture will also improve the density of the cake.

part 5 法國咖啡小點心

果仁夾心糖（牛軋糖）
Nougat blanc

法文字的 "Nougat" 牛軋早在 16 世紀就出現：我們發現時是在中世紀時期，"Nougat" 是普羅旺斯的孩子也是太陽之意。這個字也是在南法的一個地區蒙特利馬爾 Montélimar 的方言用字。在當地是一種糖漬果仁跟有蜂蜜組成的蛋糕，日後也成為這裏的特產。

在 16 世紀時，一位甜點師傅將杏仁果跟核仁果混合作為內餡，到了 18 世紀開始加入了蜂蜜跟糖，接着 20 世紀開始就出現有加入葡糖糖到煮開的糖漿裏。

最早製作出現這道甜點是在 16 世紀南法的馬賽 Marseillaise 中心，因為許多地區也認為是他們地區的特產，最後讓大家終於承認這個甜點是為蒙特利馬爾 Montélimar 這個城市的特色甜點。

在 17 世紀由一位希臘人從亞洲地區帶來一顆杏樹，因此杏樹就開始在南法開始蓬勃的生長，這位希臘人也開始製作了 13 種甜點，其中這個 Nougat 牛軋蛋糕的部份，經過這位希臘人的再度改良，變成：由蛋白、蜂蜜糖混合成主體，這個甜點在當地一年只有製作兩次，一次是復活節另一次當然就是聖誕節了！

> 窗外一片潔白的雪地，跟桌上的果仁夾心糖一樣的雪白，聖誕節的傍晚，微弱陽光照射在雪地上，玻璃窗上飄附着些微小點的白雪，冷空氣的外面對應着屋內正燒着壁爐，空氣中瀰漫着壁爐裏正燒着木材散發的香氣，那種香味好似果仁夾心糖裏的杏仁果一樣的芳香。

{ **Ingredients**

(About 700-800 grams of candy)

250 grams Granulated Sugar

300 grams Lavender Honey

80 ml Water

3 Each Egg white

100 grams Almond

50 grams Pistachio

50 grams Pine Nuts

食材（約 700~800 克糖果）

細砂糖 250 克、薰衣草蜂蜜 300 克、水 80ml、蛋白 3 顆、杏仁 100 克、開心果 50 克、松子 50 克

 製作

step 1

糖跟水放在一個深鍋裏，放在火爐上煮滾，滾後關小火讓他慢慢煮（最好是煮滾後保持在 143℃ 溫度繼續慢滾煮）煮糖的過程中千萬不要攪拌。

Bring water and sugar to a boil. Turn heat down and continue to keep it boiling at 143℃. Do not stir while cooking the sugar.

step 3

當兩個鍋子的糖與蜂蜜正在火爐上煮時候，例外準備一個鋼鍋打發蛋白，打到打蛋器挖起，蛋白霜時，蛋白霜不會滴落就可以了。

When the sugar and the honey are cooking, place egg white in a stainless steel mixing bowl. Beat with electric mixer until stiff peaks formed.

step 5

再將混合好的蜂蜜跟糖漿慢慢的倒入打發的蛋白霜裏，一邊倒一邊用打蛋器持續打發，不要停。打到蛋白霜呈現光澤且油亮狀態。

Slowly add honey sugar mixture into beaten egg white. Continue to beat with whisk while pouring. Do not stop and beat until meringue is oily and bright.

step 2

在另一個鍋子裏放入蜂蜜小火煮滾。

Bring honey to a boil in another saucepan.

step 4

將煮滾的蜂蜜倒入煮滾的糖漿裏，用木匙攪拌。

Add boiling honey into boiling syrup. Mix with wooden spoon.

step 6

把鋼鍋挪動到一個鍋裏，鍋裏加水，使用隔水加熱的方式，讓蛋白糖霜更加緊實。使用木匙一邊攪拌蛋白糖霜，這個部份比較難判斷。另一種判斷方式：原本加糖後打發的蛋白糖霜的份量在隔水加熱後的份量會變得比之前來得少，當你在攪拌蛋白糖漿時候會發現攪拌的力氣越來越會有重量感，這樣的狀態就表示已經是完成了。

Move mixing bowl on top of a saucepan. Add water into saucepan. Using the waterbath method, heat up meringue to make it tighter. Continue stirring with wooden spoon.This is quite difficult to judge. Another method for judging is: With the water bath, the volume of the meringue will decrease. When you feel that the meringue has become heavier, it reaches reachedthat stage. The process is complete.

step 7

將果仁放在不沾鍋上乾煎熟。

Stir fry nuts on non stick pan without oil.

step 8

果仁倒入蛋白糖霜裏，攪拌均勻。

Add nuts to the meringue. Mix well.

step 9

鐵盤上放上一張烘焙紙或是防油紙，把剛剛攪拌的蛋白糖果仁倒進鐵盤上，抹平整。放進冰箱冷藏至少要 24 小時，讓蛋白糖果仁變硬。

Line stainless steel pan with parchment paper. Spread meringue evenly and refrigerate at least 24 hours. This will allow the nougat to turn hard.

step 10

糖果變硬後，在使用大的刀子切成小塊狀，享用。

Once it has become hard, cut it into small pieces and serve.

Helpful Tips
亮亮的小建議

1. 煮糖漿的時候，要非常小心，糖漿的溫度非常高，被糖漿燙到是會非常非常痛，因此，煮糖漿時盡量不要用任何東西攪拌它，只要拿起鍋子搖晃讓糖漿煮均勻就好了。

2. 薰衣草蜂蜜可以更換成玫瑰蜂蜜，因為這是南法的特色甜點，我們大多用南法特產的薰衣草蜂蜜來製做。當然，一般的蜂蜜也是可以拿來製作使用的。果仁的部份：可以使用乾燥水果或是糖漬水果來製作。如果你喜歡有香料的香氣在糖果裏面，你也可以增加進去做成屬於自己特色的果仁夾心糖。

3. 在步驟 9 將蛋白雙果仁到入鐵盤抹平食的速度要快，因為糖遇冷空氣硬得很快，會抹不動。

4. 夏天做這個甜點時要注意空氣中的溫度，太熱糖果要硬化的速度較慢，需要用冷藏來讓糖果變硬。

1. *Use extra care when cooking the syrup as the temperature is very high. It will really hurt if you get burnt. Therefore, you should avoid using anything to stir the syrup while it is cooking. Just shake the pan so the syrup is even.*

2. *Lavender Honey can be replaced with rose honey. Since this is a southern specialty, we usually use its product, lavender honey, to make it. Regular honey can also be used. You can also use dried fruit and candied fruits to replace the nuts. If you want to improve the aroma of the candies, you can also add essence into the mixture for specialty candies.*

3. *On step 9, it is necessary to work fast when you pour candy mixture onto pan. You have to spread it out really quickly before it hardens.*

4. *You have to be very careful about the temperature during summer time. When it is too hot, it will take a lot longer to harden the candies. Refrigeration may be necessary.*

杏仁馬卡宏
小脆餅
··············
Les macarons
aux amandes

這 這是法國阿爾薩斯省洛林地區的特色小點心，口感十分的接近巴基斯坦的一種十字餅乾，這種餅乾在義大利非常有名。在威尼斯這種餅乾叫做 macarons，即是薄餅的意思。來到法國之後便很快的被喜愛甜點的法國人接受了。也就這樣，變成許多法國的小城鎮非常特別的甜點之一，尤其再法國北邊跟阿爾薩斯省會將這小餅乾當做一年一度的聖誕節餐後佐咖啡的小甜點來享用。

烤好的餅乾，妹妹們細心的放進她收藏已久的復古餅乾盒裏，想要細細慢慢的品嚐上一段時間。

Ingredients (About 15)

3 Each Egg White

1 dash Sea Salt

125 grams Almond Flour

125 grams Sugar

食材（約 15 片）

蛋白 3 顆、海鹽 1 撮、杏仁粉 125 克、
糖 125 克

 製作

 step 1

烤箱以 180℃ 預熱 10 分鐘。

Preheat oven to 180℃ for 10 minutes.

step 3

加入杏仁粉跟糖，攪拌到成稠狀。

Add almond flour and sugar. Mix until it is thick.

step 4

烤盤上放上一張烘焙紙 抹上少
許奶油，可以用湯匙，將蛋白杏
仁糖糊，一顆顆的放上烘焙紙
上，也可以使用擠花嘴，擠上一
個個圓形在烘焙紙上。

Line baking sheet with parchment paper. Brush lightly with oil. You can use soup spoon to scoop batter onto baking sheet. Piping bag can be used to pipe out circles onto the parchment paper.

step 2

沙拉碗裏放進蛋白跟海鹽，使用
電動打蛋器將蛋白打發。

Add egg white and salt to mixing bowl. Beat with electric mixer until stiff peaks formed.

step 5

放進烤箱烤 20~25 分鐘即可。

Bake 20-25 minutes.

費南雪
Financier

最早在 17 世紀時期由洛林省一間教會裏的修女姊妹 Nancy 製作，最主要是以杏仁粉，麵粉，糖跟蛋白混合成麵團製作為主的橢圓形小點心。這個小點心一開始的名字是「聖會修女」。基於在中世紀時期，大家沒有太過多的肉類可享食用，加上那時期大家會取用蛋黃當作教堂畫作的塗料，剩下的蛋白就拿來做這道美味的小甜點。

法國下午時刻，涼涼的樹陰下閱讀着法國文學小說「鐘樓怪人」，突然想起費南雪的來由……。

Ingredients

(Makes about 12 pieces)

75 Gram Egg white

87 grams Icing sugar

75 grams Unsalted butter

50 grams All Purpose Flour

40 grams Almond Flour

5 grams Green Tea Powder

Dash Sea Salt (optional)

5 grams Sesame (optional)

食材（約 12 顆份量）

蛋白 75 克、糖霜 87 克、無鹽奶油 75 克、麵粉 50 克、杏仁粉 40 克、綠茶粉 5 克、海鹽（不是必須的）1 撮、芝麻（不是必須的）5 克

製作

step 1

烤箱以 180℃ 預熱，混合蛋白，糖霜，過篩麵粉，杏仁粉，綠茶粉使用刮棒混合拌勻。

Preheat oven to 180℃ . Mix egg white, sugar, sifted flour, almond flour, green tea powder together with scraper.

step 2

在一個小鍋裏融化奶油直到奶油顏色呈榛果色，離開火爐，等待降溫變成手可以摸的溫度，再加入剛剛混合好的麵糊裏。

Melt butter until it turns into a hazelnut color. Remove from heat. Let temperature reduce until comfortable to touch. Add into batter.

step 3

烤模裏塗上奶油，倒入麵糊。

Brush baking mold with butter. Pour batter in.

step 4

烤約 15~20 分鐘，出烤爐 放涼後，脫膜即可。

Bake for 15-20 minutes. Let cool and remove from mold.

Helpful Tips
亮亮的小建議

1. 可以在麵糊上撒上些許海鹽幫助提味。

   Adding dash of sea salt to batter can enhance the flavor.

2. 依照自己的喜好撒上杏仁片或是芝麻。

   Sprinkle almond slices or sesame based on your liking.

125

甜漬甜橙條 &
甜漬甜橙巧克力條
Ecorces d'oranges confites

> 每季的水果盛產，吃不完的時候，我們最愛拿來製作果醬，或是水果軟糖，當然，還有甜漬水果⋯⋯自然，健康的水果零食，家人喝咖啡或茶最愛的小點心⋯⋯。

在歐洲古代，羅馬人發明用蜂蜜醃漬方式保存水果。在十字軍東征那個時期，才開始有水果軟糖、果醬跟糖漬水果的出現。這些使用各種方式來努力保存盛產時產量過剩產過多的水果。這種保存水果的方式在法國中世紀末期就變的更為盛行了。

{ **Ingredients**

100 grams Sweet Orange Peel
(about 2 oranges)

As Needed Granulated Sugar

As Needed Water

Dash Sea Salt

{ **Chocolate Sauce**

50 grams Chocolate

食材
甜橙皮（約2顆）100克、細砂糖適量、水適量、海鹽一小撮

巧克力醬
巧克力 50 克

step 3

取出甜橙皮，用刀子將黃色皮上的白膜刮掉。

Take out orange peel. Use knife to remove white part of skin.

step 1

將甜橙去皮，可以將白色薄膜部份一起切下。

Peel skin off orange. Remove white membrane also.

step 5

從新加入與甜橙皮同樣重量的水跟細砂糖到鍋子裏，煮到糖水快乾掉，但是，不能煮到變成焦糖色。（小提醒：煮甜橙糖漿的糖跟水比例以甜橙皮的重量為參考基準 1：1：1，例如：甜橙皮 100 克，水 100 克，糖 100 克，若是甜橙皮份量增加，糖跟水的比例也要跟着增加。）

Add same amount of water and sugar as orange peel into pan. Cook until syrup almost dries up. But do not cook until it turns brown. (Little reminder: The proportion of sugar, water and sweet orange peel is as follows: 1:1:1. For example, Sweet orange peel 100 grams, Water 100 grams, Sugar 100 grams. If the weight of sweet orange peel increases, sugar and water portions have to increase too.)

step 2

切下的甜橙皮跟一小撮海鹽，一起放進深鍋裏，加入水只要蓋到甜橙皮即可，煮 15 分鐘。

Add a dash of sea salt to orange peel and add into saucepan. Add water to cover peel. Cook for 15 minutes.

step 4

再度放入水裏煮 10 分鐘，倒掉水。再從新加入水只要煮 5 分鐘。甜橙皮切成你想要的長條狀。

Cook in water again for 10 minutes. Discard water. Add water and cook another 5 minutes. Cut sweet orange peel to your desired lengths.

 法國咖啡小點心

step 7

隔天，準備一盤細砂糖，將瀝乾的甜橙皮沾取糖後，在繼續曬約1小時就可以收到罐子裏了。

2 days later, prepare a pan of granulated sugar. Dip dry sweet orange peel in sugar. Continue to dry for 1 hour and then store in covered tin.

step 6

把剛剛在糖水裏煮的甜橙皮，莢出來，放在室溫內瀝乾糖漿一晚。

Remove sweet orange skin soaked in syrup. Let dry in room temperature for one night.

糖漬甜橙巧克力棒
Candied Sweet Orange Chocolate Stick

step 1

準備 50 克的黑巧克力，用隔水加熱的方式將巧克力融化。

Prepare 50 grams of dark chocolate. Use water bath method to melt chocolate.

step 2

取已經瀝乾的甜橙棒沾取巧克力後，等待巧克力乾了之後 就可以收到罐子裏，慢慢品嚐了

Dip dry sweet orange peel in chocolate. Once chocolate hardens, stick can be stored in cans for enjoyment.

Helpful Tips
亮亮的小建議

1. 做這個甜點建議最好使用自然栽種的或是有機種植的甜橙較好。
2. 切剩下來的甜橙果肉可以拿來製作蛋糕使用，本書裏有兩種甜橙蛋糕的做法，可以作為參考，這樣就比較不浪費材料。
3. 一次可以製作多份量，做兩種不同口味，煮甜橙皮的糖漿也不要丟棄，可以泡個紅茶包，加甜橙糖漿進去一塊飲用，或是加入咖啡裏也很有風味。

1. *It is best to use natural grown and organic oranges for this recipe.*
2. *Leftover orange flesh can be used for cakes. There are two orange cakes recipes in this book you can review. This way the ingredients will not be wasted.*
3. *You can do a few batches and more than one flavor at one time. Do not discard the sweet orange syrup. It can be used to add onto tea or coffee.*

{ **Ingredients**

(About 30 Pieces of cookies)

150 grams All Purpose Flour

2 Tbsp Cocoa Powder

75 grams Dark Brown Sugar

100 grams Room Temperature Softened
Butter

2 Each Egg Yolks

A Handful Shelled Pistachios (unsalted)

Dash Sea Salt

食材（約 **30** 塊小餅乾）

麵粉 150 克、可可粉 2 湯匙、黑糖 75 克、室溫融化奶油 100 克、蛋黃 2 顆、
去殼開心果（沒有調味的）一把、海鹽一小撮

130

step 1

在一個沙拉碗裏將奶油，糖，海鹽電動打蛋器打混合均勻。麵粉＆可可粉過篩後加入，混合攪拌均勻。接着加入兩顆蛋黃，這時候盡量可以用手將麵團揉均勻。

Beat butter, sugar, sea salt with electric mixer until combined. Add sifted flour and cocoa powder. Mix. Then add 2 egg yolks. Mix dough with hand at this time.

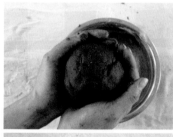

step 2

將巧克力麵團揉勻後，接着揉成光滑球狀。保鮮膜包起放進冰箱約一小時或是室溫放一小時亦可。

Form dough into a smooth ball. Cover with plastic wrap and refrigerate for 1 hour or let it sit in room temperature for an hour.

step 3

烤箱以 200℃ 預熱 10 分鐘。

Preheat oven to 200℃ for 10 minutes.

step 4

拿出巧克力麵團，工作台上撒上些麵粉，將麵團擀平約 5mm 厚度，用餅乾模壓出餅乾形狀，烤盤上放上一張烘焙紙，放上壓出模型的餅乾，在放上一顆開心果。

Take out chocolate dough. Sprinkle flour onto work area. Roll dough out to 5 mm thickness. Use cookie cutters to cut out several shapes. Line baking sheet with parchment paper. Place out cut dough onto baking sheet. Put a pistachio on top of each cookie.

step 5

進烤箱烤約 12~15 分鐘，出爐放涼即可享用。

Bake for 12-15 minutes. Cool and serve.

Helpful Tips
亮亮的小建議

1. 此餅乾跟松露巧克力一樣，可以做有加海鹽口味，也可以做辣椒粉口味。這些調味對巧克力都有提味與增加層次風味的效果。

2. 做好的餅乾可以堆疊起來，用漂亮的緞帶或是繩子綁起來，放進袋子裏送人也很大方。

1. This cookie is just like the chocolate truffles. You can add sea salt or pepper flavours to it. These flavours will enhance the taste of the chocolate.

2. Finished cookies can be stacked together. They can be wrapped with pretty ribbons and strings, and bagged as gifts.

在 法國東南邊最早先的一個小鎮上有一位專門製作糖果的商人，
Louis Dufour 路易士·居福 先生，最早發明了"巧克力泥漿"
這個甜點。在 1985 年的有一天，他在自己的甜點廚房裏開始他第
一次用簡單且簡便的方式製作出這道令人醉心的甜點，這個甜點便
成為他在 1985 年末最令他開心也是令他感到美好的一件事。他偉大
又簡單的想法是：將可可與鮮奶油再加上一些些的香草香氣，將它們攪
拌混合在一起，最後，將已經變成巧克力泥的圓球包裹着薄薄的巧克力粉之
後，呈現在大家的面前。看到完成品的大家當時覺得這個甜點看起來似乎真的很不錯。Louis
Dufour 路易士·居福先生發現大家的接受度很高，也就變成他最驕傲的一個事蹟，並且在法
國甜點歷史界上成為一件很成功的甜點發明事件。之後，法國美食家將這道美味的甜點改名為
"松露巧克力"，由於它的外型長得與松露非常相似，口感也跟松露一樣的美味，因而得此名。

松露
巧克力
Les truffes au
chocolat

這個稍有點繁複卻又傳統做法的甜點（相較於市面上許多的松露巧克力做法來做比較），
是我第一次用這麼貼近的方式來揭開他傳統手法的神秘面紗，今晚，全家很慎重的準備
盛上一杯好的法國白蘭地跟香檳酒就是要來搭配這道甜點……因為他只會出現在盛大場
合跟節慶日上例如聖誕節或是跨年。

Ingredients (About 25 pieces)

250 grams Dark Chocolate

2 Tbsp Milk

75 grams Room Temperature Butter

1 Tbsp Heavy Whipping Cream

1 Tbsp Brandy

1 Tsp Vanilla Essence

Dipping Ingredients

150 grams Dark Chocolate

50 grams Cocoa Powder

食材（約 25 顆）

黑巧克力 250 克、牛奶 2 湯匙、室內融化奶油 75 克、濃縮鮮奶油 1 湯匙、白蘭地 1 湯匙、香草精 1 茶匙

完成沾黏食材

黑巧克力 150 克、可可粉 50 克

製作

step 1

巧克力壓成碎片，與牛奶放進鍋裏，用隔水加熱的方式來融化巧克力。

Chop chocolate into small pieces. Add milk to chocolate and melt over water bath.

step 2

取起巧克力鍋，慢慢加入一小塊一小塊的奶油，並混合在一起。混合完之後加入濃縮鮮奶油，香草精，跟白蘭地酒，混合後，倒入一個小盤子或碗，放進冰箱讓巧克力泥漿變硬。

Slowly add chunks of butter into melted chocolate. Mix together. Once combined, add heavy whipping cream, vanilla essence and brandy. Pour mixture into small tray or bowl. Refrigerate until chocolate mixture turns hard.

step 3

當巧克力泥漿變硬後，使用湯匙將變硬的巧克力泥漿挖放在手上，整成圓形，在放進盤子裏，放進冰箱，等待一會兒使用。

When chocolate mixture becomes hard, use tablespoon to scoop chocolate onto hand. Shape into balls. Put balls onto tray and refrigerate for later use.

step 4

將 150 克的巧克力，用隔水加熱方式再度融化。融化之後，從冰箱拿出剛剛整形好的巧克力泥漿，一顆顆的放進融化的巧克力岩漿裏，用兩根叉子做輔助將巧克力岩漿裏上巧克力泥漿球上，在放在有可可粉的盤子裏，沾取可可粉後，放進另一個盤子裏。

Melt 150 grams of chocolate using the water bath method . Once the chocolate is melted, take the chocolate balls out from the fridge. Dip balls into melted chocolate. Use two sticks or chopsticks to help coat chocolate balls in melted chocolate. Then roll coated balls with cocoa powder. Place them all on a different tray.

step 5

所有的巧克力泥漿球沾取完可可粉後 可以放進罐子裏，放進冰箱冷藏……隨時享用。

Once all the chocolate balls are done, they can be put into tins and refrigerate until consumption.

椰子
剛果球
Congolais

椰子剛果球主要以椰子絲跟糖與蛋白混合後可以做成圓形球或是三角形，也可以用擠花嘴擠出有花樣的球狀，低溫烤之後就可以享用了。法國家庭裏的奶奶們喜歡一次烤很多，裝進餅乾盒或是密封罐裏，等下午茶時，泡杯熱熱的茶一邊喝茶一邊享用。現在的法國甜點麵包店盛至大型賣場都有賣一包包的椰子剛果球，但是，還是沒有自己做得來得好吃！

風和日麗，微風徐徐吹來的下午，跟朋友一塊輕鬆製作甜而不膩的剛果椰子球，朋友笑說，這個不油膩的甜點很適合正在減肥的她⋯⋯在家裏庭院裏享受女生們的瘦身午茶⋯⋯。

Ingredients (About 12 pieces)

1 Each Egg White

63 grams Shredded Coconut or coconut Powder

50 grams Granulated Sugar

5 grams Vanilla Candies

食材（約 12 顆）

蛋白 1 顆、椰子絲或是椰子粉 63 克、
細砂糖 50 克、香草糖 5 克

製作

step 1

烤箱以 160℃ 預熱 10 分鐘。

Preheat oven to 160℃ for 10 minutes.

step 2

在一個沙拉碗裏放進糖，椰子粉
和香草糖用手混合均勻。

Mix sugar, coconut and vanilla candies by hand.

Helpful Tips
亮亮的小建議

一次可以烤多一些，一部份的椰
子絲剛果球可放進密封的小罐
子裏，放進冰箱，下回姊妹們聚
會下午茶可以在拿出來享用。

You can bake a bigger batch at one time. Part of the Congolais can be put into airtight container back to the fridge for future use.

step 3

另一個沙拉碗裏放入蛋白 加入
一些的海鹽，使用電動打蛋器打
成硬性蛋白。（小提醒：硬性蛋
白就是蛋白會變成像雪一樣堅固
的狀態，可以用攪拌棒挖起來之
後，蛋白被打發後都不會掉下來
的狀態就是 ok 的囉！）

In another bowl, beat egg white and sea salt with electric mixer until stiff peaks formed. (Little Reminder: Stiff Peak egg whites is when egg whites are stiff as snow. Even when stirred with whisk, it will not fall apart)

step 4

將硬性蛋白加入糖＆跟椰子粉混
合在沙拉碗裏，用刮棒將它們攪
拌混合均勻，一定要攪拌到不能
是液態狀，也不可以過乾喔！烤
盤上放上一張烘焙紙，將椰子蛋
白糊用糖匙挖起來在手中塑成你
想要的形狀（小提醒：椰子蛋白
糊太液態狀或是過乾都不易塑造
形狀喔）

Slowly fold sugar coconut mixture into egg white. It cannot be liquidy or too dry. Line baking sheet with parchment paper. Mold coconut meringue into desired shapes. (Little Reminder: if the meringue is too wet or too dry, it will be hard to shape)

step 5

160℃ 烤 12 分鐘，烤出爐後放涼
即可。

Bake at 160℃ oven for 12 minutes. Cool and serve.

瑪

德蓮是來自皇室的其中一項甜點，故事的起源在 1755 年法國皇帝路易十五的岳父 Stanislas Leszczynski（原波蘭國王），後來被封為洛林公爵，在 Commercy 省的城堡裏舉辦一場大型宴會，在用餐的期間，一名僕人小心地走近公爵，悄悄地告訴他，城堡裏的甜點師傅在用餐時間離開了城堡，並且將宴會即將要享用的甜點一併帶走了。

為了讓宴會能夠順利完成，一位名為叫 Madeleine Paulier 的僕人提出建議，她可以製作她奶奶的拿手甜點來做為晚宴最後的餐後甜點。公爵為了不使這個突如其來的事件影響到宴會的進行，便答應此項建議。沒想到，宴會上大家品嚐享用到此甜點後大感驚訝，紛紛詢問這道美味甜點的名字，公爵便順水推舟將這道小點心稱為「Madeleine 瑪德琳」，此後，Madeleine 小點心變成 Commercy 鎮上的著名點心。

趁着陽光充足的下午，在廚房一邊聽着法國爵士樂一邊製作鬆軟的法國小點心，可以依照自己的心情喜好加入想要的口味或是香氣在鬆軟迷人的 Madeleine 裏……鬆軟綿密的 Madeleine 沾着熱茶一起享用，香氣讓人不小心一口接着一口的吃光，連我家的柴犬 Lucky 也等着吃。

瑪德琳
Madeleine

{ **Ingredients** (About 24 pieces)

150 grams All Purpose Flour

126 grams Room Temperature Butter

150 grams Sugar

2 Each Eggs

2 Tbsp Milk

1/2 Tsp Baking Powder

1/2 Tsp Vanilla Essence

食材（約 24 顆）

麵粉 150 克、室溫融化奶油 126 克、糖 150 克、蛋 2 顆、牛奶 2 湯匙、泡打粉 1/2 茶匙、香草精 1/2 茶匙

step 1

將蛋跟糖放入沙拉碗裏打混，讓顏色略成淡白色（小提醒：蛋跟糖打混後，糖會融化在蛋裏，顏色就會比先前蛋的顏色在淡一些喔！）

Beat eggs and sugar until slightly pale. (Little Reminder: When eggs are mixed with sugar, sugar will dissolve in the eggs, causing the color of the eggs to lighten.)

step 2

加入麵粉跟牛奶，泡打粉，香草精攪拌混合後，加入奶油在攪拌成霜狀麵糊，放進冰箱約 30 分鐘（小提醒：麵粉＆泡打粉最好一起過篩後加入，攪拌時才可以避免結塊喔！）

Add flour, milk, baking powder and vanilla essence. Mix well. Next add in butter and beat until thick and fluffy. Refrigerate for 30 minutes (Little reminder: It is best for flour and baking powder to be sifted together and add to batter. This will prevent any lumps)

step 3

烤箱以 200℃ 預熱 5 分鐘。烤模塗上奶油跟撒上少少的麵粉。（小提醒：塗奶油跟撒麵粉在烤模上會比較好脫膜喔！）

Preheat oven to 200℃ for 5 minutes. Brush baking mold with butter and sprinkle with a little flour. (Little Reminder: This will enable easy release)

step 4

取出冰箱的麵糊，放進約 7~8 分滿的麵糊。

Take dough out from fridge and fill molds 70 to 80% full.

step 5

送進烤箱 200 ℃ 烤 3~4 分鐘，降低 180℃ 烤 6~8 分鐘。

Bake at 200℃ oven for 3-4 minutes. Reduce temperature to 180℃ and bake for another 6-8 minutes.

Helpful Tips 亮亮的小建議

1. 如果瑪德琳在烤箱裏烤上色太過快，可以直接降低溫度烤上色，不需要等待烤完 3~4 分鐘再降低溫度喔！

2. 如果想增加瑪德蓮的香氣，可以加入甜橙或是檸檬絲進入麵糊裏攪拌後，放進冰箱。這樣的瑪德琳有自然的水果香氣喔！為了飲食安全健康，請務必使用有機不噴灑農藥的柑橘類水果喔。

1. If the madeleines brown too quickly, you can reduce the temperature before the 3-4 minutes is up.

2. If you want to improve the aroma of the madeleines, you can add shredded oranges or lemon to the batter before putting it into the fridge. This way the madeleines will have natural fruit fragrance. For safety reasons, it is best to use citrus without pesticides.

列日 華夫餅

Gaufres liege

列日華夫餅可以是餅塊狀或者是像薄薄的格子脆餅。法國北邊以往的新年期間，他們稱呼它為"上帝的地方"或是"上帝的恩賜"。法北早期過年的時候，人們會準備很多列日華夫餅送給窮人或是弱勢家庭，作為他們的新年禮物，另外的含義是代表新的一年希望能替這些窮人家或是人們帶來更多好運與新希望。這些列日華夫餅後來變成老饕與熱愛甜食的法北人會在當有客人來訪時，搭配咖啡或是酒一起享用的小點心。

阿嬤跟阿嬤的媽媽各有自己不同的做法，這兩種我都享用過。

提供這道列日華夫餅的食譜是隔壁鄰居阿嬤私房做法，跟我家阿嬤的食譜比較起來，做法要在簡單一些，口感上多了蘭姆酒香與肉桂香；搭配一杯溫溫熱熱的紅酒應該算是法國鄉下人詮釋愜意人生的另一種表現法。

> 隔壁阿嬤說最喜歡冬天吃著這餅拿著杯香料溫紅酒，看著窗外凋落的樹葉，手上的溫紅酒的香氣跟餅乾的肉桂香氣互相呼應著，讓身體跟心一起溫暖了起來。

Ingredients (About 12 pieces)

250 grams All Purpose Flour

125 grams Brown Sugar

125 grams Butter

1 Each Egg

50 grams Rum

Dash Cinnamon Powder

1 Bag Vanillla Candies

食材（約 12 片）

麵粉 250 克、紅糖 125 克、
奶油 125 克、蛋 1 顆、蘭姆酒 50 克、
肉桂粉一撮、香草糖 1 包

製作

step 1

麵粉倒進一個盤碗裏。

Pour flour into mixing bowl.

step 2

麵粉中間挖出一個洞放進紅糖，
全蛋、融化奶油，肉桂粉，蘭姆
酒跟香草糖。

*Make a hole in the middle of the flour
and add sugar, egg, melted butter,
cinnamon powder, rum and vanilla
candies.*

step 3

混合所有的材料，攪拌成麵糊，
在揉成圓形後，放進冰箱，約
1~2 小時。

*Mix all ingredients into a batter and
form into a round circle. Refrigerate for
1-2 hours.*

step 4

拿出麵團，揉成小圓型麵團。放
上鬆餅機上，約 5 分鐘即可拿
出。直到所有的麵團壓烤完畢。

*Take dough out and mold into little
round balls. Put them onto waffle
machine. Cook them for 5 minutes until
all dough has been used.*

part 6

法國甜點
基本練功夫

香脆塔皮
pâté de sablée

香 脆塔皮和酥塔派皮主要以奶油跟麵粉混合製作而成的。香
脆塔皮也可以稱為又稱為"甜塔皮",除了主要的奶油,
麵粉之外還有糖跟蛋,也有一些甜塔皮的做法是不放蛋以牛奶來
取代蛋的做法。

"基本塔皮"跟法國的"奶霜醬"都是法國甜點裏最基本組合甜點最
主要的元素,在法國甜點變化萬千的食譜裏最主要的變化基礎都是從這
兩種基礎甜點基底來發展到創意甜點。近幾年法國的許多甜點店裏的甜塔皮&奶霜也都開始
發展出自家獨特食譜做法,也都是必須先從基本功夫練起,才進而發展擁有自己獨特配方的
獨特味道的塔皮。

甜塔皮在法國甜點的用途十分廣泛,舉凡在所有法國甜點相關書籍出現的"甜派"或是"塔"
都幾乎是用香脆甜塔皮來做的,例如:草莓塔、藍莓塔、蘋果派、巧克力派、反烤蘋果派……
都是。

一開始,製作塔皮可能會有些許不習慣。如果不習慣用手揉,家裏若有揉麵機也可以用揉麵
機 來幫忙。手揉的好處是當手揉時加以觀察揉塔皮時與過程中間變化的差異性,可能前一兩
次會失敗,但是,我相信只要按照步驟一步步慢慢做,你一定可以做出與法國甜點店一樣好
吃的塔皮,甚至會更好。

Ingredients (For a 12 inch Tart)

300 grams Medium Gluten Flour

100 grams Butter

125 grams Icing Sugar

1 Each Egg

食材（**12 寸塔大小**）

中筋麵粉 300 克、奶油 100 克、
糖霜 125 克、蛋 1 顆

製作 -

step 1

麵粉與糖霜一起過篩後，放在桌
上或是沙拉碗裏，用手稍微攪拌
混合。

*Sift flour and icing sugar. Put them on
the working area or mixing bowl and
slightly mix them with hands.*

step 2

加入奶油塊，用手將奶油跟麵粉
一起捏混合，最後的狀態是麵粉
有點成麵包屑的狀態。

*Rub butter into flour mixture with
hands. Batter with be crumbly.*

step 3

奶油麵粉中間挖個洞，打入一顆
全蛋，將蛋跟奶油麵粉和勻揉成
光滑麵團狀。

*Make a hole in the center of the batter,
add an egg. Knead batter into a smooth
dough.*

step 4

保鮮膜包起來放進冷藏約 30 分鐘。

Wrap dough with plastic wrap. Refrigerate for about 30 minutes.

step 5

麵團從冰箱拿出，工作台上撒上些許麵粉，擀麵棍亦是。將麵團擀成 0.3~0.5 公分厚度，將麵皮用擀麵棍捲起，挪到塔模上。

Remove dough from fridge, sprinkle flour on working area and rolling pin. Roll dough to about 0.3 to 0.5 mm thickness. Roll dough up with rolling pin and bring it to tart mold.

step 6

將派皮貼緊塔模後，在使用擀麵棍在塔模邊緣將多餘的麵皮去掉，用手再將塔模頂邊邊的派皮整形整好。蛋白霜不會滴落就可以了。

Fit pie crust tightly onto mold. Remove any excess with rolling pin. Fix dough around the rim so it fits snugly onto mold.

step 7

放進冷藏約 30 分鐘。用叉子，在塔皮底部插出洞口，放上石鎮，烤箱以 200℃ 預熱 10 分鐘，烤 10~15 分鐘即可。

Refrigerate for 30 minutes. Prick holes on bottom of crust with fork. Put in stone weight. Preheat oven to 200℃ for 10 minutes. Bake. 10-15 minutes.

step 8

烤出爐後，拿出石鎮，塔皮放涼等待備用。

Once it is done, remove stone weight. Cool and use.

Helpful Tips

亮亮的小建議

1. 在第 7 步驟裏，可以在塔皮底部插出洞之後，放上一張烘焙紙在放上一堆石鎮，若是沒有石鎮可以用任何豆類果實壓着派皮進去烤箱烤。（請注意：烤過的豆類請勿再來拿來食。）

2. 塔皮上戳洞是為了在烤的過程中，讓派裏產生的氣可以由孔排出，才不會塔皮一直往上膨脹，會產生塔皮內沒有足夠的空間可以填入餡料。

3. 混入麵粉的奶油請先切小塊後，再放入麵粉裏用雙手戳散時比較方便使用。

4. 奶油請盡量使用剛從冰箱取出的奶油來製作派皮。因為，奶油會因為手的溫度關係，慢慢的軟化，所以，使用剛從冰箱取出有點硬的奶油會比較好跟麵粉搓揉混合再一起。

5. 麵團整型到最後成光滑狀時，雙手跟工作台上應該就不會有麵團屑了，因為麵團會將工作台跟你雙手的麵團都吸收了！也就是整形好麵團時工作台＆你的雙手會是乾淨的。

6. 派皮一次做多一些，包好放在冷凍庫可以放 3 個月，冷藏只能一週，在冷凍庫取出需要退凍約 40~50 分鐘，在推擀開來。

7. 烤派皮剩下的零碎派皮可以再度整形擀開用餅乾膜壓出模型，直接當成烤餅乾給小朋友沾果醬或是巧克力醬一起吃。

1. In step 7, after pricking holes in the bottom, you can place a parchment paper on top and put in the stone weights. If stone weights are not available, you can replace it with beans and nuts. (Please note that you should not use baked beans for consumption)

2. Pricking holes in the pie crust allows it to let the air out Therefore it will not continue to expand, thus leaving enough room for the fillings.

3. Please dice the butter before adding into the flour mixture. It is more convenient when you are rubbing them into the flour with your fingers

4. Use chilled butter for best results. Butter will slowly melt with the warmth of the fingers. Therefore, chilled butter will be easier for mixing.

5. Once dough is smooth, there should not be any residual dough on the working area. The finished dough should have absorbed all the bits and pieces from the working area and your hands. Therefore the working area and your hands will be very clean.

6. Your can make a few batches of the pie crust and they will keep well in the freezer for 3 months. However, refrigeration can only keep it for 1 week. You will need to defrost dough for 40-50 minutes when taking out from freezer before it can be rolled out and used.

7. Leftover scraps can be put together and shaped with cookie cutters to make cookies for children. They are good with jam and chocolate spread.

酥塔皮
pâte brisée

{ **Ingredients** (For 12 inch tart)

250 grams Medium Gluten Flour

120 grams Butter

5 grams Sea Salt

60-90 grams Water

食材（約 12 寸塔大小）

中筋麵粉 250 克、奶油 120 克、海鹽 5 克、水 60~90 克

酥塔皮也可以稱為又稱為鹹派皮，除了主要的奶油，麵粉之外還有海鹽跟水，也有一些鹹派皮的做法是不放水以牛奶來取代的鹹塔皮。也有一些新的做法會多加入蛋來增加鹹派皮的風味。

鹹塔皮在法國甜點的用途主要是在製作法式鹹派，舉凡在所有法國甜點相關書籍出現的"鹹派"或是"鹹塔"都幾乎是用酥甜塔皮來做的，例如：洛林鹹派，馬鈴薯培根鹹派，鮭魚菠菜茴香鹹派……都是。但是，也有些水果派會用鹹派皮，例如：蘋果派。這是一些法國甜點師傅新創意的做法，或許你下回也可以嘗試看看。

亮亮的小建議

酥塔皮＆香脆塔皮的建議事項都是相同的，更詳細的注意跟建議事項請見香脆塔皮。

Same as Pate de Sablee

製作

step 1

麵粉過篩，加入海鹽用手將混有海鹽的麵粉攪拌一下。

Sift flour, add sea salt and mix with hands.

step 2

加入切小丁的奶油，用手將麵粉與奶油搓揉成麵包屑狀。

Add diced butter. Rub them into batter until crumbly.

step 4

最後用雙手將麵團揉成光滑球狀，保鮮膜包起放冰箱 30 分鐘。

Finally, knead mixture into smooth dough. Wrap with plastic wrap for 30 minutes. Refrigerate.

step 3

奶油麵粉中間挖個洞，加入水，從水中間劃圈圈方式慢慢混合水跟麵粉。

Make a hole in the center of the batter, add water. Slowly run hand in circles to mix flour with water.

step 5

冰箱拿出後，擀開的方式＆烤的方式與 香脆塔皮（甜塔皮）步驟 5~9 都是相同的！

Remove from fridge. Roll out and bake like sweet tart crust. Steps 5-9 are the same!

千層派皮在法國的甜點界有着舉足輕重的地位，以厚重的奶油跟麵粉為主，少許的海鹽跟和水，多次的桿擀、摺之後，展現出有如薄如紙般一層一層派皮，由於有着麵粉包住厚重奶油，口感多為香脆，奶油香味濃郁。

千層派皮
pâte feuilletée

千層派皮的運用跟香脆派皮＆酥塔派皮相比下有更廣泛的用途，例如，派、修頌、國王餅……也會運用在一些派塔上或一些開胃前菜的部份，變化性相當的多。但，卻也是非常花時間製作的派皮。缺點是：天氣不能太热，除了奶油容易融化，在擀派皮的過程中，會因為天氣熱，奶油融化造成派皮有點溼，在擀派皮時候容易有破洞。

因此在擀千層派皮的時候，你會很需要低溫且涼爽的氣溫。另外，手的溫度過高也是另一個容易造成揉擀派皮的過程中使讓麵派皮太容易過軟、過溼的另一項重要因素（擀香脆派皮＆酥塔派皮亦是）。手溫過熱的解決方式是：最好將雙手放進冰箱或是冷凍庫冷卻降溫，或浸冰冷的水也是可以的，手出冰箱或是冰水後立即將雙手抹去水份在進行揉派皮動作，並且在揉的過程當中，雙手要不斷的撒上乾麵粉，讓手的溫度不至於會沾黏麵粉或是奶油。千層派皮跟其他兩種派皮一樣，擀好後分切用保鮮膜包好，放進保冷袋或是保冷盒裏，可以保存3個月。要使用的時候在從冷凍庫取出退冰到正常的狀態，便可以開始進行擀平派皮的動作！派皮的顏色如果跟當初擀製的時候有落差，如：顏色有變深，表示派皮有變質，新鮮期過了。建議丟棄不宜使用，應重新再擀製一份新的。

千層派皮的擀製方式發展到現在有複雜度高，等待時間久的，也有複雜程度較低，等待時間較短的。建議大家可以先試試我的這個屬於簡易方式的做法，熟悉與漸漸有手感後可以在進一步練習複雜度較高的千層派皮作法。

Ingredients

(Simple Layered Pie Crust)

500 grams Medium Gluten Flour

10 grams Sea Salt

250 grams Water'

375 grams Butter

食材（簡易程序千層派皮）

中筋麵粉 500 克、海鹽 10 克、 水 250 克、奶油 375 克

製作

step 1

麵粉過篩，麵粉中間挖個洞放進海鹽，並且加入四分之三的水。

Sift flour . Make hole in center and put in sea salt and 3/4 water.

step 2

將麵團揉混合，不要摺疊派皮。加入剩下的水，再將麵粉跟水全部揉混合均勻。

Mix dough together. Do not fold pie dough. Add remaining water. Mix into dough.

step 3

麵團揉成圓球狀，保鮮膜包起來放冰箱約 20~30 分鐘（請將冰箱的冷度調至最冷）。

Shape dough into round ball. Cover with plastic wrap and refrigerate for 20-30 minutes. (Adjust fridge temperature to the coolest).

step 4

從冰箱拿出麵團，用刀子再圓麵團上切十字，在用手撥開。稍加擀平約 30×30cm 寬跟長度。

Remove dough from fridge,. Cut a cross with knife on top of dough. Move dough out with hands into cross. Roll dough out to 30×30cm.

step 5

平均的將奶油放進呈十字狀派皮的中間，將派皮由下往上摺，由上往下摺，由左往右摺，再由右往左摺。將摺好的四個小角落黏好，不能有洞口，摺好後的派皮約 25×25cm 大小。（請注意：派皮一開始摺的方向要記好，接下來的五次擀摺的方向都要一致一樣，不能錯亂，錯亂會影響到派皮烤好後的口感喔。）

Evenly put the diced butter in the middle of the dough. Fold pie dough from bottom to top, top to bottom, left to right, then right to left. Secure 4 corners. Do not leave any holes. Folded pie dough should be about 25×25 cm. (Reminder: You have to remember the direction of folding when first started. The next five folding sequence have to be the same and cannot go wrong. If the sequence is mixed up, it will affect the texture of the crust).

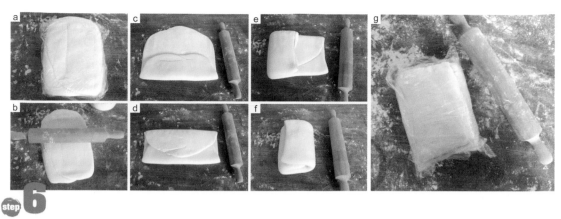

step 6

再步驟 5 的最後摺疊是由右往左摺，表示開口在左邊，因此，每次從冰箱拿出來時要將開口朝左邊，再開始擀。擀至約 45×35cm 再繼續將派皮由下往上摺，由上往下摺，由左往右摺，再由右往左摺，以保鮮膜包起放進冰箱 20 分鐘。

As the last folding on step 5 is from right to left, opening should therefore be on the left. Therefore, when the dough is taken from the fridge each time, the opening will face the left . Continue to roll. Roll out to 45×35 cm. Continue to fold dough from bottomto top, top to bottom, left to right and right to left, Cover with plastic wrap and chill in thefridge for 20 minutes.

step 7

從冰箱拿出將開口朝左邊開始擀平派皮，中間過程可以撒少許的麵粉避免沾黏。再開始步驟 5 和 6 的動作兩次後，派皮以保鮮膜包起放進冰箱再冰 20 分鐘。

Take dough out from fridge, with opening facing left, roll dough again. You can adda little flour if it gets sticky. Repeat Step 5 and Step 6 twice. Cover in plastic wrap andrefrigerate for another 20 minutes.

step 8

從冰箱拿出將開口朝左邊開始擀平派皮，將派皮由下往上摺，由上往下摺，由左往右摺，再由右往左摺這是擀摺第 5 次，也是最後一次。擀摺好後，用尺量好千層派皮的寬度。可以分切成三份。

Remove dough from fridge again, with opening facing left, roll dough. Fold dough frombottom to top, top to bottom, left to right, right to left. This is the fifth foldingand the last folding. After the folding is done, measure the width of the pie crust. Dividedough into three servings.

step 9

上圖是兩份，在分別用保鮮膜包起來放進保冷袋 / 盒收進冷凍庫就完成了。

The picture above shows two servings. Wrap separately with plastic wraps and storeinto freezer.

Helpful Tips
亮亮的小建議

1. 一般複雜程序的千層派皮做法，在氣溫很高的亞洲製作，大約要 8 個小時才會完成。這個簡易方式操作時間約 6 個小時，即可以完成。

2. 如果室內氣溫真的很熱，建議最好是擀摺一次就放進冰箱，放冰箱的時間也要跟著拉長拉到 45 分鐘再拿出來擀。這會耗時長一些時間，但是派皮比較不會破。派皮破了會不好處理，擀麵棍跟工作台沾的都是奶油，接著會黏到派皮上，造成派皮破損度變高。

1. *For complicated method of layered pastries, in high temperature Asia, it will take about 8 hours. Using the simplified method ,it will take about 6 hours.*

2. *If room temperature is very high, a suggestion is chilling the dough after every folding. The chilling time has to be proportionally longer at 45 minutes . Even though it is a longer process, it will ensure that the pie crust will not break. If crust breaks, it will be hard to handle. Rolling pin and working area will be greasy and when it gets to the pie crust, it can be very damaging.*

戚風蛋糕
Génoise

戚風蛋糕是法國許多蛋糕體的基本款，都是以戚風蛋糕作為主體，在發展其他變化的創意蛋糕。

主要是以蛋、糖跟麵粉組成的蛋糕。做法也有許多的不同。我家的小朋友超愛戚風蛋糕，每次烤好一小塊的戚風蛋糕，小朋友們就會搶着拿去塗果醬、巧克力醬……就開始吃的津津有味。

如果，你想要替家人或是你的朋友做個慶生蛋糕，那麼就做一個戚風蛋糕吧！將蛋糕橫切成兩塊，夾層裏面放進果醬，或是法式香堤在放上自己喜歡的水果，簡單美味的蛋糕就馬上呈現了，不需要花很多時間製作。相信吃過你自己做的戚風生日蛋糕你的家人們或是朋友們一定會對你做的蛋糕讚不絕口。

戚風蛋糕可以改變成其他口味，例如：巧克力、香草……等。大部份都以原味的戚風蛋糕做為蛋糕的主體比較多，若要變化口味或是增加蛋糕整體的風味，會以奶油或是水果來做表現方式。另外，也可以在蛋糕完成最好後，塗上法式香堤醬或是巧克力醬（例如微多麗雅蛋糕、巧克力蛋糕……書中都有示範做法可以參考）讓你的手作蛋糕變成更佳完整，美麗呦！

{ **Ingredients** (Serves 6)

4 Each Eggs

120 grams Granulated Sugar

120 grams All Purpose Flour

食材（約 6 人份）

蛋 4 顆、細砂糖 120 克、麵粉 120 克

製作

step 1

烤箱以 200℃ 預熱 10 分鐘。

Preheat oven to 200℃ for 10 minutes.

step 2

準備一個沙拉碗,將蛋黃跟蛋白分開,蛋白放進沙拉碗裏,使用電動打蛋器打發後,將 120 克的細砂糖慢慢的加入,請記得要一邊加糖一邊繼續打混蛋白跟糖。

Separate egg whites and egg yolks. Place egg whites in mixing bowl and beat with electric mixer. Gradually add 120 grams of sugar. Make sure you do not stop beating when sugar is being added.

step 3

糖都依序加完後,放入蛋黃繼續攪拌打混合均勻。

After all sugar has been added, add in egg yolks and continue to mix batter.

step 4

將麵粉過篩後,分批加入剛剛攪拌均勻的蛋糖糊裏,使用刮棒慢慢將麵粉跟蛋糖糊攪拌混合。

Sift flour and add it by batches onto the batter. Use spatula to slowly fold it into the batter.

step 5

烤膜裏塗上奶油,剪下一張烘焙紙放進烤模裏後,倒入麵糊,進烤箱烤 25 分鐘。

Brush baking pan with butter. Cut a parchment paper and fit it into the pan. Pour in batter. Bake for 25 minutes.

step 6

出烤爐後,放涼約 2 分鐘,在用刀子的尖部在烤模周圍小心的劃一圈,方便讓戚風蛋糕倒扣脫膜。

After cake is done, cool it for 2 minutes. Use a sharp knife and run it around the cake pan. This is allow easy release of the cake.

Helpful Tips
亮亮的小建議

1. 如果不用烘焙紙放在烤膜裏,可以在烤膜的底部跟周圍塗上奶油,在撒上薄薄的麵粉,這樣也可以方便蛋糕脫膜。

2. 若是不想在烤模上塗上奶油建議剪一塊與烤模底部相同大小的烘焙紙,在剪一塊比烤模高度要在高出 5 公分的烘焙紙貼在烤膜四周,接著在倒入麵糊,這樣也可以方便蛋糕脫膜。

3. 測試蛋糕裏面有沒有熟成的方法:取一跟長牙籤插入蛋糕中心部位後取出,如果牙籤表面上有點溼黏表示蛋糕還沒烤熟,相反的,牙籤取出,表面上是乾的,就可以立即將蛋糕取出。

4. 如果要將蛋糕橫切成兩塊,最好等待蛋糕涼了之後在切,請記得要用有鋸齒的刀子切蛋糕,才不會造成切蛋糕時,蛋糕塌陷喔!

5. 麵糊裏如果有太多氣泡,可以用牙籤或是叉子稍微戳破,這樣烤出來的蛋糕就不會太多氣孔了。

1. If you do not put parchment paper in pan, you can lightly brush the bottom and sides of the pan with butter then sprinkle with flour. This will also help with the release.

2. If you do not want to brush pan with butter, cut parchment paper the size of the bottom and sides of the pan and fit them into the pan.

3. To test if the cake is done, use a long toothpick and stick in into the middle of the cake. If the toothpick comes out wet, it means the cake is not cooked yet. On the other hand, if it comes out dry, the cake is done.

4. If you want to slice the cake in half, please let it cool down first. Make sure you use a serrated knife. Otherwise the cake will collapse.

5. If there are too many bubbles in the batter, you can break them with a toothpick. This way the finished cake will not have too many air bubbles.

法國甜點基本練功夫

Ingredients (Approx 250 grams)

200 grams Fresh Cream

20 grams Icing Sugar

食材（約 250 克份量）

鮮奶油 200 克、糖霜 20 克

法式香堤
Crème chantilly

製作

step.

將鮮奶油放進沙拉碗裏，使用電動打蛋器先以低速打發鮮奶油，鮮奶油開始起泡泡之後，慢慢分次放進糖霜，並將電動打蛋器切換到高速將鮮奶油糖霜打到成霜狀即可。

Whip fresh cream with electric mixer at low speed. Once you start to see bubbles on the cream, slowly and gradually add in the sugar. Beat the cream mixture on high until it is thick and creamy.

Helpful Tips

亮亮的小建議

1. 天氣如果太熱，可以事先將沙拉碗、電動打蛋器的攪拌棒與鮮奶油放進冰箱冷藏約 30 分鐘後在打，比較容易成功。

2. 請先將電動打蛋器放進裝有鮮奶油的盆裏再啟動電動打蛋器，同樣地，關閉電動打蛋器時也必須攪拌棒還在盆內就關閉，避免鮮奶油噴得自己滿臉或是噴灑四處。

3. 打好的法式香堤可以做很多用途使用，例如蛋糕裏的夾層奶霜或是蛋糕外層的奶霜塗抹，或是放在熔岩巧克力上一起享用，搭配水果果醬＆新鮮水果一起享用也很棒。

1. *If the weather is too hot, you can chill the mixing bowl and whips from electric mixer for 30 minutes before use. This will make it easier to get better results.*

2. *Make sure you turn on the electric mixer only after you put it into the mixing bowl. Same thing applies when it was turned off. Otherwise the cream will splatter all over your face and everywhere.*

3. *There are many uses of the Chantilly cream. You can use it to ice cakes, use will melted chocolate or accompany jams and fresh fruits.*

Ingredients (Approx 250ml)

2 Each Egg Yolks

250 ml Milk

1/2 Each Vanilla Bean

50 grams Granulated Sugar

食材（約 250ml）

蛋黃 2 顆、牛奶 250ml、
香草莢半顆、細砂糖 50 克

英式香草淋醬

Crème anglaise

英 式香草醬是由牛奶、蛋黃、糖做成的液態淋醬，在加入香草莢
來增加醬汁的香氣。在英國，最有名的用法會將淋醬煮得較為
濃稠，並且要熱熱的享用。

英式香草淋醬，可以單獨被使用，最廣泛的被用在與"漂浮島"（是
一種蛋白打發的甜點，將醬汁盛放在盤子下方，蛋白漂浮島則放在醬
汁上方）一併享用。

相同的，英式香草淋醬，在法國也會被用在各種蛋糕上，當作蛋糕的甜
醬汁，或是水果的淋醬。另外一個比較廣泛的用法則是淋在熔岩巧克力上，
跟被挖開的流出猶如岩漿般的巧克力岩漿混合一起享用。

在法國，我們使用習慣使用香草莢在淋醬裏，而且將冰涼的英式香草淋醬淋上喜歡的甜點上
享用啊！

製作

牛奶放進深鍋裏，香草莢對切，
用刀尖刮出香草籽，放入鍋內以
小火煮滾。

*Add milk and scraped vanilla seeds into
saucepan. Use low heat and bring to a boil.*

step 3

倒入小火煮滾的香草牛奶。

*Pour vanilla milk into the egg sugar
mixture.*

step 5

放涼後，放冰箱約 3 小時候即可
使用。

*Cool and chill for 3 hours. It is now
ready for use.*

Helpful Tips

亮亮的小建議

1. 煮好的醬汁會有些許的泡
泡，放置冰箱後，就會慢
慢的消退了！

2. 如果沒有香草莢，可以以
1/2 茶匙的香草精來替代。

1. *Cooked sauce will have some
bubbles. Once it is cooled, the
bubbles will disappear.*

2. *If you do not have vanilla beans,
you can use 1/2 tsp of vanilla
essence instead.*

step 2

準備另一個沙拉碗，放入蛋黃跟
糖攪拌均勻。

Mix egg yolks and sugar in a bowl.

step 4

再將混合已經牛奶香草跟蛋黃糖
過篩後倒回深鍋裏，在火爐上煮
約 3 分鐘略稠即可。

*Run mixture over sieve and bring it back
to the heat. Cook it for about 3 minutes
until slightly thickened.*

法國甜點基本練功夫

151

巧克力醬
Chocolat fudge

巧 克力醬是一種平常又基本，且常在法國許多樣甜點裏出現的甜醬。

最主要必須使用超過 70% 以上 Cacao 可可的巧克力，接着是融化巧克力火的溫度。必須要小火慢慢融，火太大、溫度過高，容易發生巧克力跟油分離的狀態。融好的巧克力要有點光澤感，才算是好的成品。巧克力醬通常使用在塗抹蛋糕夾層裏，也可以當作蛋糕外層裝飾包裹着蛋糕體。例如有：聖誕節的巧克力樹輪蛋糕，經典的濃郁巧克力蛋糕，單獨使用可以是巧克力生膏，或是製作巧克力生膏塔。另外也會被運用在杯子蛋糕上的裝飾巧克力醬。蛋糕上的裝飾……作用很廣泛。

Ingredients (About 500 grams)

200 grams Dark Chocolate (chopped)

30 grams Brown Sugar

200 grams Diced butter

2.5 Tbsp Extended Shelf Life Milk

1/2 Tsp Vanilla Essence

食材（約 500 克）

黑巧克力，敲小塊 200 克、
紅砂糖 30 克、奶油，切小塊狀 200 克、
保久乳 2.5 湯匙、香草精 1/2 茶匙

製作

step 1

將所有的材料放進鍋裏，慢火將
鍋內巧克力跟奶油融化，使用木
匙攪拌，讓所有的材料混合融合
在一起。

*Put all ingredients into saucepan. Melt
butter and chocolate over low heat. Stir
and combine everything with wooden
spoon.*

step 2

融好的巧克力醬倒入另一個碗裏。

*Pour melted chocolate fudge into another
bowl.*

step 3

放入冰箱約一小時後使用。

*Cool and chill for 1 hours it is now ready
for use.*

Helpful Tips
亮亮的小建議

1. 如果沒有紅砂糖，可以改用白細砂糖。但是，紅砂糖比較沒有那麼
   甜，細砂糖比較甜。

2. 牛奶部份：請盡量使用保久乳，經過高溫殺菌的保久乳比較穩定。

3. 巧克力的部份請用超過 70% 以上成份的可可來製作。

1. *If you have no brown sugar, use caster sugar instead. Please be reminded that brown
   sugar is less sweetness than caster sugar.*

2. *For the milk, you use pasteurized milk is better. Quality of the milk is stable because of
   being sterilized.*

3. *Recommend to use 70% dark chocolate.*

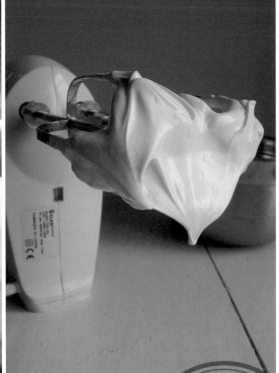

法式
蛋白霜
Meringuée

法 國蛋白霜主要以蛋白和糖兩種基本元素組合而成。在打發蛋白
之前可以放進一撮海鹽來打發，蛋白比較容易成形。法式蛋
白霜最常出現在檸檬塔上面，現在許多的檸檬塔上方的蛋白霜都是
用義式蛋白霜的做法。法式蛋白霜與義式蛋白霜的差別在何處呢？
法式蛋白霜打發後需要送進烤箱，以低溫烘烤成硬脆口感。義式蛋白
霜則是以高溫水滾的方法，隔水加入打發蛋白，再加入加熱的糖水繼續
打發到蛋白霜非常濃稠，厚重感。義式蛋白霜因為隔水加溫的關係，已經生
的蛋白經過間接高溫殺菌，因此，打發的義式蛋白霜擠在塔上後，可以直接用噴火槍燒出焦
糖色後直接享用。

法式蛋白霜還可以擠出各種造型或大塊，進烤箱低溫烘烤後，塗上果醬享用，變成另一種蛋
糕的形式出現。小的蛋白霜餅乾有些會加入食用色素，製作出美麗、優雅、吸引人顏色的蛋
白霜脆餅。近來的法國甜點裏，出現許多將蛋白霜擠成大水滴狀烘烤後，在蛋糕上做各種好
看時尚的造型，或是放在巧克力飲品上方。除了令甜品的視覺美觀用處以外，還是另類的法
式甜點享用法。

Ingredients

(150 grams of meringue)

40 grams Icing Sugar

1 Each Egg White

Dash Sea Salt

食材（150 克蛋白霜）

糖霜 40 克、蛋白 1 顆、海鹽一撮

製作

step 1

鋼鍋裏放進蛋白跟海鹽。電動打蛋器低速打發。

Beat egg white and salt with electric mixer at low speed.

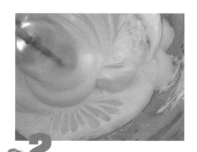

step 2

鍋裏的蛋白出現許多大顆的泡泡時，加入一部份的糖，將電動打蛋器切成高速繼續打，大顆泡泡開始不見，蛋白開始越來越緊密。

When you start seeing large bubbles, gradually add in part of the sugar. Continueto beat at high speed until the large bubbles disappear and the egg white becomesmore dense.

step 3

將剩下的糖加入，直到打到蛋白泡沫變得更細緻。

Add remaining sugar, and beat until egg white is smooth.

step 4

只要將蛋白霜用刮棒挖起來時，蛋白泡沫很堅固，不會掉下來即可了！

Egg white will still stay firm when being lifted by spatula. It is then ready.

step 5

若要烤成"蛋白脆餅"可以上述的材料加少許的檸檬汁一起打發，之後。裝進擠花袋裏，擠在烤盤上用 100℃烤 60 分鐘即可。

If you want to make meringue cookies, you can add a little lemon juice and beat it together. Put batter into piping bag and squeeze into baking pan. Bake at 100℃ for 60 minutes.

step 6

可愛的小蛋白霜脆餅，還可以放在熱巧克力或是咖啡上搭配一起飲用。

Sweet mini meringue cookies can also be put into hot chocolate and coffee.

Helpful Tips

亮亮的小建議

1. 蛋白請選用冰的蛋白比較容易打發。如果蛋白一直在室溫要多加 20 克的糖來打，較容易打的發。

2. 打過頭的蛋白霜會有類似棉花狀也就是分離狀，不扎實，沒有彈性，也不細緻，更不容易消泡，這樣的蛋白霜烤起來不會膨脹。就是失敗品了。成功的蛋白霜要有彈性，紮實，細緻而且不易流動。

3. 打蛋白霜時候，加入糖開始要一直使用高速打發，約打 25~30 分鐘左右。當然份量越多打得時間也會越久。

4. 烤好的蛋白脆餅怕潮濕，遇濕的空氣，原本脆硬的蛋白脆餅很容易變黏手。製作好要快速裝進罐子裏。

1. *Chilled egg white is easier to whip up. If the egg white is rest at room temperature, 20g of sugar more should be added.*

2. *If beaten egg white has a cotton ball look, then it is falling apart and not firm. It is not elastic and smooth. The meringue will not expand when baked and will fail. Successful meringue has to be elastic, firm, smooth and does not slide around.*

3. *Once sugar is added to the egg white, you have to turn the mixer on high speed. Beat it for about 25-30 minutes. Time increases as quantity increases.*

4. *After baked meringue, store in a tin in case of the meringue meets moist air to be made sticky on the hands.*

附錄

食材名詞對照表

法文	香港	台灣	英文
Farine	麵粉 / 餅粉	麵粉	Flour / Cake Flour
Poudre d'Amande	杏仁粉	杏仁粉	Almond Powder
Maïzena	粟米粉 / 粟粉	玉米粉	Corn Flour / Corn Starch
Sucre en poudre	幼砂糖 / 糖粉	細沙糖	Fine Sugar
Sucre glace	糖霜	糖霜（粉）	Icing Sugar
Sucre perlé	珍珠糖	糖晶	Pearl Sugar
Sucre roux	紅糖	紅細紗糖	Brown Sugar
Cassonade	黃糖 / 咖啡糖	蔗糖	Cane Sugar
Miel	蜜糖	蜂蜜	Honey
Levure fraîche	新鮮酵母	新鮮酵母菌	Fresh Yeast
Levure boulanger	乾酵母 / 酵母	乾燥酵母粉	Dry / Instant Yeast
levure chimique	發粉 / 泡打粉	泡打粉	Baking Powder
bicarbonate de soude	梳打粉 / 小蘇打	蘇打粉	Baking Soda
Sel fin	海鹽 / 幼海鹽	細海鹽	Fine Sea Salt
Oeuf	雞蛋	蛋	Egg
Jaune d'oeuf	蛋黃	蛋黃	Egg Yolk
blanc d'oeuf	蛋白	蛋白	Egg White
Lait	牛奶	牛奶	Milk
Beurre	牛油	奶油 / 黃油	Butter
Crème fraîche liquide	淡忌廉 / 忌廉	液態鮮奶油	Whipping Cream
Crème fraîche épaisse	凝脂奶油 / 濃厚奶油	濃縮鮮奶油	Heavy Cream
Riz rond	圓米 / 短米	圓米	Round Rice
Chocolat	朱古力	巧克力	Chocolate
Chocolat noir	黑朱古力	黑巧克力	Dark Chocolate
Poudre en cacao	朱古力粉 / 哈咕粉 / 可可粉	可可粉	Cocoa Powder
Café expresso	濃縮咖啡 / 特濃咖啡	濃縮咖啡	Expresso
Gousse vanille	雲喱拿枝 / 雲喱拿豆	香草莢	Vanilla Pod
Sucre vanille	雲喱拿糖	香草糖	Vanilla Sugar
Cannelle	肉桂	肉桂	Cinnamon
Aneth	刁草 / 蒔蘿	茴香	Dill
Rhum	冧酒 / 蘭姆酒	蘭姆酒	Rum
Cidre	蘋果酒	蘋果酒	Cider
Cognac	白蘭地 / 拔蘭地	白蘭地	Brandy
Vin blanc sucré	甜白酒 / 白葡萄酒	甜白酒	Sweet White Wine
Raisin sec	提子乾 / 葡萄乾	葡萄乾	Rasin
Orange	橙	甜橙 / 橘子 / 柳橙	Orange
Zeste d'orange	橙皮	甜橙皮 / 橘子皮	Orange Zest
Fraises	士多啤梨 / 草莓	草莓	Stawberry
Sucre confit	蜜餞水果	糖漬水果	Candied Fruit
Citron	檸檬	檸檬	Lemon
Zeste citron	檸檬皮	檸檬皮	Lemon Zest
Rhubarbe	大黃	大黃根	Rhubarb

法文	香港	台灣	英文
Pomme	蘋果	蘋果	Apple
Myrtilles	藍莓	藍莓	Blueberry
Poires	梨	西洋梨	Pear
Prune	西梅 / 布冧	黑李	Plum / Prune
Carotte	紅蘿蔔 / 胡蘿蔔 / 甘筍	紅蘿蔔	Carrot
Poudre noix de coco	椰子粉	椰子粉	Coconut Powder
Noix de coco râpe	椰絲	椰子絲	Grated Coconut / Desiccated Coconut / Shredded Coconut
Fromage blanc	白芝士	白乳酪	Fromage Cheese
Amande	杏仁	杏仁果	Almond
Pistache	開心果	開心果	Pistachio
Pignons pin	松子仁	松子果	Pine Nut
Noisette	榛子	夏威夷果	Hazelnut
Figues sèche	無花果乾	無花果乾	Dried Fig
Noix	核桃 / 合桃	核桃	Walnut

容器換算 & 烤箱溫度

在做甜點時，在混亂的時刻有時候我們可能臨時找不到容器或是沒有磅秤可以秤重，這種情況真的很令人苦惱。因此，以下我提供基本常在使用的容量換算的方式跟最簡單的容器使用容量轉換表。即使沒有專業或是齊全的工具在家做甜點仍然可以輕鬆自在的做。

關於液體牛奶，水，酒……

1L 公升 =	1000ml（毫升）
	10dl（分升）
	100cl（厘升）

關於 1 茶匙（咖啡匙亦可）轉換其他材料重量

1 茶匙 =	5g 海鹽，奶油或是糖霜
	4g 麵粉，油或是細砂糖
	3g 研磨胡椒粉，可可粉或是玉米粉

關於重量適用於所有的材料與有重量的材料

1Kg 公斤 =	1000g（公克）

關於 1 湯匙轉換其他材料重量

1 湯匙 =	3 茶匙或是 3 咖啡匙
	15ml 或是 1.5cl
	12g 麵粉、玉米粉
	5g 海鹽、細砂糖、奶油或是鮮奶油
	18g 米
	20g 糖晶或是粗海鹽

關於烤箱溫度換算

小火	1	30℃
	2	60℃
中小火	3	90℃
	4	120℃
	5	150℃
中火	6	180℃
大火（快熱）	7	210℃
	8	240℃
超大火（超快熱）	9	270℃
	10	300℃

國家圖書館出版品預行編目（CIP）資料

法國甜點家中出爐 / 陳芋亮編著. -- 初版. --
臺北市：橘子文化，2014.08
面；　公分
ISBN 978-986-364-021-9（平裝）

1. 點心食譜

427.16　　　　　　　　103014465

法國甜點・家中出爐

編　　著　陳芋亮
策　　劃　 ARTSONA
翻　　譯　Monica Kwan
編　　輯　郭麗眉
設　　計　萬里機構製作部

出 版 者　橘子文化事業有限公司
　　　　　萬里機構出版有限公司　聯合出版
總 代 理　三友圖書有限公司
地　　址　106台北市安和路2段213號4樓
電　　話　（02）2377-4155
傳　　真　（02）2377-4355
E-mail　　service@sanyau.com.tw
郵政劃撥　05844889　三友圖書有限公司

總 經 銷　大和書報圖書股份有限公司
地　　址　新北市新莊區五工五路2號
電　　話　（02）8990-2588
傳　　真　（02）2299-7900

SAN YAU
http://www.ju-zi.com.tw
三友圖書
友直 友諒 友多聞

初　　版　2014年八月
定　　價　新臺幣380元
　ISBN　　978-986-364-021-9（平裝）

地址： _____ 縣/市 _____ 鄉/鎮/市/區 _____ 路/街

_____ 段 _____ 巷 _____ 弄 _____ 號 _____ 樓

廣 告 回 函
台 北 郵 局 登 記 證
台北廣字第2780號

SANYAU

三友圖書有限公司　收

SANYAU PUBLISHING CO., LTD.

106　台北市安和路2段213號4樓

SANYAU

三友圖書／讀者俱樂部

填妥本問卷，並寄回，即可成為三友圖書會員。
我們將優先提供相關優惠活動訊息給您。

優質好康

粉絲招募
歡迎加入

◦ 看書 所有出版品應有盡有
◦ 分享 與作者最直接的交談
◦ 資訊 好書特惠馬上就知道

旗林文化╳橘子文化╳四塊玉文化
https://www.facebook.com/comehomelife

親愛的讀者：
感謝您購買《法國甜點・家中出爐》一書，為感謝您的支持與愛護，只要填妥本回函，
並寄回本社，即可成為三友圖書會員，將定時提供新書資訊及各種優惠給您。

1 您從何處購得本書？
□博客來網路書店 □金石堂網路書店 □誠品網路書店 □其他網路書店
□實體書店＿＿＿＿

2 您從何處得知本書？
□廣播媒體 □臉書 □朋友推薦 □博客來網路書店 □金石堂網路書店
□誠品網路書店 □其他網路書店＿＿＿＿□實體書店＿＿＿＿

3 您購買本書的因素有哪些？(可複選)
□作者 □內容 □圖片 □版面編排 □其他＿＿＿＿

4 您覺得本書的封面設計如何？
□非常滿意 □滿意 □普通 □很差 □其他＿＿＿＿

5 非常感謝您購買此書，您還對哪些主題有興趣？(可複選)
□中西食譜 □點心烘焙 □飲品類 □瘦身美容 □手作DIY
□養生保健 □兩性關係 □心靈療癒 □小說 □其他＿＿＿＿

6 您最常選擇購書的通路是以下哪一個？
□誠品實體書店 □金石堂實體書店 □博客來網路書店 □誠品網路書店
□金石堂網路書店 □PC HOME網路書店 □Costco
□其他網路書店＿＿＿＿ □其他實體書店＿＿＿＿

7 若本書出版形式為電子書，您的購買意願？
□會購買 □不一定會購買 □視價格考慮是否購買 □不會購買
□其他＿＿＿＿

8 您是否有閱讀電子書的習慣？
□有，已習慣看電子書 □偶爾會看 □沒有，不習慣看電子書
□其他＿＿＿＿

9 您認為本書尚需改進之處？以及對我們的意見？
＿＿＿＿＿＿＿＿＿＿＿＿＿＿＿＿＿＿＿＿＿＿＿＿＿＿＿＿＿＿＿＿＿＿

10 日後若有優惠訊息，您希望我們以何種方式通知您？
□電話 □E-mail □簡訊 □書面宣傳寄送至貴府 □其他＿＿＿＿

謝謝您提供寶貴的意見，
您填妥寄回後，將我們將不定期提供
最新的會訊與優惠活動資訊給您：

姓名＿＿＿＿＿＿＿＿ 出生年月日＿＿＿＿＿＿＿＿

電話＿＿＿＿＿＿＿＿ E-mail＿＿＿＿＿＿＿＿＿＿

通訊地址＿＿＿＿＿＿＿＿＿＿＿＿＿＿＿＿＿＿＿＿＿